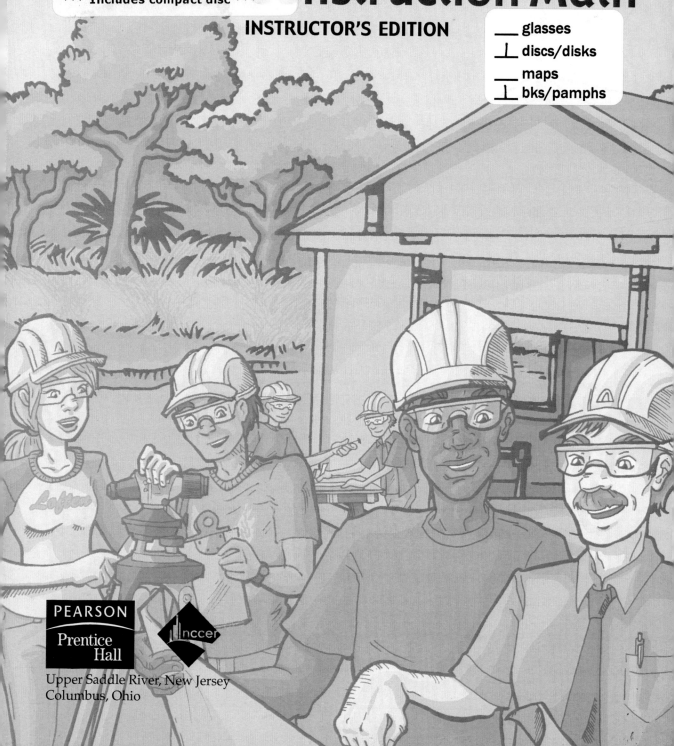

National Center for Construction Education and Research

President: Don Whyte
Director of Product Development and Revision: Daniele Stacey
Production Manager: Jessica Martin
Product Maintenance Supervisor: Debie Ness
Editors: Bethany Harvey, Brendan Coote, Carla Sly

Pearson Education, Inc. *Product Manager:* Lori Cowen

Writing and development services provided by Topaz Publications, Liverpool, New York.

Cover and chapter opener art by Jennifer Jacobs.

Copyright © 2006 by the National Center for Construction Education and Research (NCCER), Alachua, FL 32615, and published by Pearson Education, Inc., Upper Saddle River, NJ 07458. All rights reserved. Printed in the United States of America. This publication is protected by Copyright and permission should be obtained from the NCCER prior to any prohibited reproduction, storage in a retrieval system, or transmission in any form or by any means, electronic, mechanical, photocopying, recording, or likewise. For information regarding permission(s), write to: NCCER Product Development, 13614 Progress Boulevard, Alachua, FL 32615.

10 9
ISBN 0-13-227300-4

Dear Instructors:

Just seeing the word "math" strikes fear and frustration in the heart and mind of most students. Teachers also understand that a math lesson gets the same reception as any other dreaded chore: "Eat your brussels sprouts, take out the garbage, and learn your MATH."

As the former Director of the Academy of Construction Technology at Loften High School in Gainesville, Florida, this is the book I wish I had when I was teaching carpentry. Over the past two decades I have experienced the pain, embarrassment, and frustration most of my students felt because of their struggles with basic math. I feel that students must see the relevance in learning before they will invest the time and commitment needed to master the subject. Hopefully this book will help students understand the fundamentals of math in a way that is engaging, interesting, and relevant. This book's unique and real-life approach will help you demonstrate how learning and, more importantly, understanding math will help students reach their personal and professional goals.

I hope this book will be a rewarding and fun experience for you and your students. As they progress through this book, students will realize that to become successful in the construction industry, they must understand the connection between success and the ability to do basic math. Math is the language of the construction industry—success hinges on the ability to speak it well.

I would like to thank all of the students I've taught in my carpentry program over the years. They frustrated me at times, they humbled me occasionally, and they inspired me always. I hope the lessons I taught them will have a positive impact on their lives. What they taught me has made me a better educator, husband, father, and friend. I am forever in their debt.

Sincerely,

Ed Prevatt

Senior Manager of Workforce Development
National Center for Construction Education and Research

INTRODUCTION

It seemed like such a great idea and it worked. Every year for the last five years, Mr. Whyte's construction class at Lofton High School in Gainesville, Florida built a house under their teacher's guidance. The students got paid for their work and got some good experience, too. They earned money, learned a trade, and learned math in real life situations. What work the kids couldn't do was subcontracted to other construction professionals. Homebuyers liked it because they knew that Mr. Whyte supervised every job the kids did. And boy, was he tough! If it wasn't done right, it got redone.

Usually Mr. Whyte had too many potential homebuyers, but this year was different—it seemed like no one wanted them to build a house. It was a big disappointment for the kids. They all needed this project for one reason or another—some for money, some for experience, and others for the sense of accomplishment.

Al's smart and good-natured and he'd like to go to college, but he's a realist. He lives way out in the country with his grandmother and they need money. Al has his heart set on buying a red Mustang so he can get a job—all he needs is $1,000.

Phil's trying to decide whether he wants to get a construction job after high school or go to college. Without some solid construction experience, he'd have to start at the bottom, so he may as well go to college. It would make his parents happy, but Phil's not sure about it. Why bother spending another four years of his life in school when he already knows what he wants?

Tom doesn't want anyone to know this, but his dad is hard on him. Nothing Tom does seems to please him. His dad's a construction foreman and Tom desperately wants his dad to be proud of him.

Sandy knows what she wants and goes for it, but she's having a hard time this year. Sandy needs the project to keep her mind off her troubles.

Jorge just moved from San Diego, and even though all of the kids are friendly to him, he still feels like an outsider. He longs to be part of a team.

TABLE OF CONTENTS

Chapter 1 Show Me the Money

Al wants a red Mustang convertible, but how long will it take him to earn the $1,000 to buy it? Mr. Whyte and the rest of the class help Al figure it out. This chapter uses money to teach the students about decimals and division. On the way, they get a lecture on minimum wage and federal versus state laws. Topics covered in this chapter include:

- Division
- Decimals and fractions
- Solving for unknowns
- Federal and local law

Chapter 2 It's All About Space

Mr. and Mrs. Weaver want to build a house, but how much space do they need? Mr. Whyte's not around, so Phil, Tom, and Sandy help them decide. The Weavers are a nice couple, but they don't understand about surface area, so the students need to describe square units of measure and teach the couple how to calculate the area of squares, rectangles, triangles, and trapezoids. During the discussion, the Weavers learn a little about regional house designs. Topics covered in this chapter include:

- Square inches, square feet, and square yards
- Surface area of squares, triangles, rectangles, and trapezoids
- Drawing scale

Chapter 3 Where Do You Live?

Mr. Whyte teaches the class how to calculate the average cost of a home per square foot. He also explains how climate, code requirements, population, and local economics can make the same home cost a lot more in different areas of the country. Topics covered in this chapter include:

- Averages
- Exponents, powers of ten, and scientific notation
- Division
- Economics
- Building codes

Chapter 4 Cattle Country

Mr. Whyte gets a break. The Browns contract to have a house built in a rural part of Alachua County so the class is in the house building business, but there's a hitch. Mr. Whyte can't be on the building site as much as he'd like, so it's up to the class to take over. First on the agenda is a survey. The team works with a surveyor to learn about angles, direction, and area. They also learn a little about cattle because they get an unexpected visit from a neighbor's herd. Topics covered in this chapter include:

- Reading measurements
- Azimuth and bearing (directional)
- 3–4–5 rule for right triangles and the Pythagorean theorem
- Area
- Elevation
- Legal descriptions

Chapter 5 Breaking Ground

It's time to break ground and the class needs to make decisions about site preparation. They learn how to estimate the volume of soil for fill. Topics covered in this chapter include:

- Converting inches, feet, and yards
- Converting cubic inches, feet, and yards
- Volume of objects

Chapter 6 Payday

The students get their first paycheck, and Mr. Whyte can't pass up an opportunity to talk about percentages. This time he uses payroll taxes as his teaching tool. Topics covered in this chapter include:

- Converting decimal to percent and percent to decimal
- Performing calculations with percentages
- Powers of ten
- Pie charts
- Social Security, Medicare, and income taxes

Chapter 7 Divide and Conquer

There's a lot to a good foundation. The group learns about reinforced concrete. They calculate the volume of irregular shapes so they can order the concrete. Topics covered in this chapter include:

- Volume of irregular shapes
- Calculating the amount of concrete required for construction

Chapter 8 Choosing Teams

Things are really rolling now. The group is divided into carpentry, plumbing, and electrical teams. This chapter follows the rough carpentry team through math exercises. Topics covered in this chapter include:

- Linear measure
- Reading measurements
- Board feet
- Perimeter
- Estimating materials for construction

Chapter 9 Gravity Can Work for You or Against You

The plumbing team learns that drainage can be a tricky business when they become the plumber's helpers. Topics covered in this chapter include:

- Linear measure, angles, volume, pressure, grades, and slopes
- Water pressure
- Introduction to trigonometry

Chapter 10 Shocking, Simply Shocking

The electrical team has their problems, and part of it is the price of the copper wiring to get power to the house. It's a good thing one of the students turns out to be an electrical whiz. Topics covered in this chapter include:

- Calculating the circumference of a circle
- More uses for the Pythagorean theorem
- Ratios
- More about trigonometry

Chapter 11 First I'm Hot, Then I'm Cold

The teams reunite to install the heating and air conditioning system. A neighbor stops by and tells everyone what life was like in rural Florida before central heating and air. Topics covered in this chapter include:

- Volume and Btus
- Wood-burning stoves
- Celsius and Fahrenheit

Chapter 12 Inside and Out

The kids learn how to apply math skills to finish the interior and exterior of the house. Topics covered in this chapter include:

- Linear measure and surface area of flat and three-dimensional objects
- Converting between square feet and yards
- Pitch and span of roofs
- Calculating the materials needed for a job

Chapter 13 The Bottom Line

The job is almost done, and the class learns a bit about running a financially sound business. Topics covered in this chapter include:

- Calculating costs
- Determining profit

Chapter 14 Everyone Has an Angle

This chapter wraps up trigonometry functions. Topics include:

- Sine, cosine, tangent, cotangent, secant, and cosecant
- Solving unknowns

Epilogue

An epilogue is sometimes used to wrap up a story. This section tells you what the students did after graduating from high school. You might be in for a few surprises.

ABOUT THIS INSTRUCTOR'S EDITION

This book is designed to be a supplement to the regular math textbook used in a classroom. Students should use this text to help them sharpen their math skills after they have received classroom instruction. This book has been written to provide students with a transcript of what a teacher typically goes through in the classroom to solve a mathematical problem. Encourage your students to read the text and follow the calculations carefully and repeatedly until they understand the problem and the solution.

This Instructor's Edition provides teaching tips at the front of each chapter, along with solutions to the Practice and Review Problems. Additional questions can be found on the CD provided with this this book. The CD also includes helpful tables and other reference material.

ACKNOWLEDGMENTS

NCCER gratefully acknowledges the organizations and individuals that have contributed to the development of Applied Construction Math.

We would like to thank the following reviewers for their valuable feedback during the development process. Their helpful comments have made the final product a better book.

Kelly Almond, Lee County High School, GA

Steeny Chester Banks, McIntosh County Academy, GA

James A. Beal, A.V.S., Manhattan Construction Company, TX

Laura Beavers, Construction Academy Career Center, Omaha Public Schools, NE

Lori Blake, ABC Central Florida Chapter

Tammy Mose Cooper, Atlantic High School, FL

Bob Dickerson, Red Education Consulting Services, NC

Ken Eaves, Native American Vocational & Technical Education Program–Council of Athabascan Tribal Governments, AK

William E. Feher, Sussex Tech Adult Division, DE

Connie Howard, East Ridge High School Construction Career Academy, TN

Mary Jenkins, East Ridge High School Construction Career Academy, TN

Mike Mayfield, Lakeview Fort Oglethrope High School, GA

Paul McConnell, New England Institute of Technology, RI

Heather Mielke, Burlington High School Construction Career Academy, WI

Tim Spicer, Sussex Tech Adult Division, DE

Chapter 1
Show Me the Money

In Chapter 1, students need to understand that there are different ways to write the same mathematical function. Explain that the slash mark is the same as a division sign, and when the slash is used, the top number is the numerator and the bottom number is the denominator.

Students also need to understand decimal place values. In advanced classes, you may be able to relate decimal place holders to power of ten.

Many students have trouble relating decimal numbers and fractions. Since most people are very serious when money is involved, you may want to use coins to illustrate this concept. In advanced classes, you can add percentages to the discussion.

The discussion about income taxes is a good time to explain that personal income tax is the main source of government revenues. Point out to the students that when they start to work, they will pay income taxes. Then introduce any federal project—it can be local, national, or historical. Ask them if they think the project is worthwhile. Then ask them if they'd be willing to contribute cash to it. The following pie chart shows the revenue for the US government for fiscal year (FY) 2001.

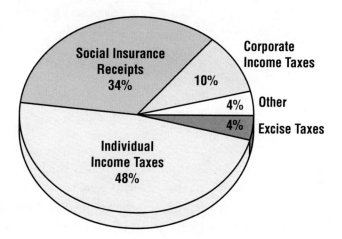

This topic can be used to stimulate discussion with the class about how our taxes are spent. Advanced classes may understand percentages, but you may prefer to use fractions on the pie chart to reinforce the lesson. The following chart shows general expenditures for 2001.

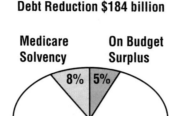

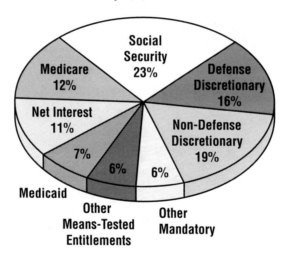

Most people believe that a sizable amount of the US budget is for foreign aid. This is not true. In 2004, less than 1% of the budget was for foreign aid. Most of the US budget is used for social programs, such as Social Security, Medicare, and Medicaid. A large amount of the budget is allocated to paying interest on the federal debt.

Chapter 1

Show Me the Money

"Okay, guys. Get in here and settle down," Mr. Whyte said as he herded the crowd of teenagers into the classroom. "Get your homework out."

"Are we going to build a house this year, Mr. Whyte?" asked Tom, who was slouched in his chair with his long legs sprawled in the aisle next to him.

"Don't know, Tom. We don't have a buyer yet, and we don't have enough money in the bank to start one on our own."

Mr. Whyte had started a house-building program about five years ago. Every year, his class worked with local building contractors to build a house. Students in his classes worked as part of the construction crew. It was popular with them because they earned minimum wage for the work they did and got some good experience. The school got part of the profits from the sale of the house, and Mr. Whyte hoped that eventually the profits would be enough to build a house even without a buyer—but that time hadn't come yet.

"Shoot," Al said, rubbing his shiny bald head. "I had my eye on some wheels."

"Maybe you can get another job, Al," Phil said.

"Man, I live way out in the country. I need wheels just to get to a job."

Al's shoulders slumped and his smooth face settled into hard lines. Mr. Whyte looked at Al and thought, he's right. Mr. Whyte knew Al lived with his grandmother in a run-down wooden house in rural Alachua County, about fifteen miles east of the city of Gainesville. Al got on the school bus every morning at six-thirty to make it to Lofton High by eight. His grandmother supported both of them on her Social Security check. Some extra cash in that household would be welcome, but Al was right. He needed transportation to get a job.

"What kind of car, Al?"

Al sat a little straighter in his chair and leaned forward. "A guy who lives near me is selling his Mustang. It's a convertible. He wants a thousand bucks. Don't guess I'll be gettin' it now."

"Why do you want a convertible, Al?" Tom asked. "You need a little pick-up truck."

Mr. Whyte laughed. "I think Al will look good in a convertible. Don't give up hope, Al. I'm talking to a couple of potential buyers tonight. If there's a way, I'll find it."

"I got faith in you, Mr. Whyte." Al's good humor was restored for the moment.

"A thousand dollars. That's a lot of money." Mr. Whyte twisted the end of his long mustache. The class let out a collective groan. Mr. Whyte was always latching onto a math problem—any math problem. "Anyone know how much the minimum wage is here in Florida?"

"I think it's $5.15. I read it off a poster at Wal-Mart," Jorge said.

"That's the federal minimum wage, Jorge. Florida's higher now." Mr. Whyte used the Spanish pronunciation for Jorge.

"How can it be higher, Mr. Whyte?" Phil asked. "We learned in Government class that when the federal government makes a law, all the states need to follow it even if they don't want to."

– 1.2 –

"That's right, Phil. The states do need to follow federal law, and Florida does. People get at least $5.15 an hour, but do you remember when people were out in front of Hitchcock's supermarket last year getting signatures on petitions? That was to get a proposition on the ballot to raise the state's minimum wage so we could vote on it. States can make laws that are stricter than federal laws, so now Florida's minimum wage is $6.15 an hour, and it's going to go up again next year."

Mr. Whyte walked up the whiteboard. "Back to Al's car. We know two bits of information. Al will earn $6.15 per hour and he needs $1,000. Now, what else do we need?"

"We need to know how many hours he's going to work a week," a skinny kid with dyed black hair said.

"Not quite, Travis. We need to know the total number of hours Al needs to work." Mr. Whyte wrote on the whiteboard. "$1,000 divided by $6.15 equals the number of hours worked. Some of you might not have thought about money this way, but the cents part of $6.15 is a decimal number. Anyone know why?"

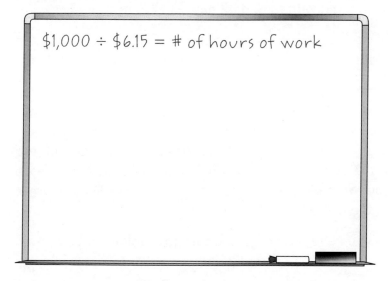

"Yeah, 'cause it's after the decimal point," Al said.

"Nice try, Al," Mr. Whyte said. "But let's be more specific. Anyone else?"

"Because fifteen cents is only part of a dollar," Jorge said.

"Good. Cents are less than a dollar, so decimals are less than one." Mr. Whyte wrote this on the whiteboard.

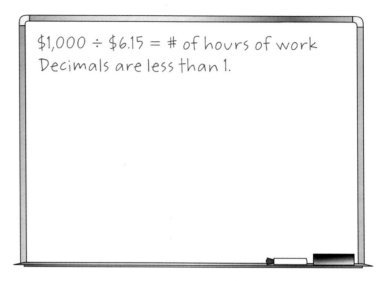

"So what units are pennies in? Olivia?" Mr. Whyte pointed to a girl with long dark hair.

Olivia leaned her head to the left, thinking. "I guess a hundred, since there are a hundred pennies in a dollar."

"You almost got it. A penny is a hundredth of a dollar. That's the units for when something is divided into a hundred parts. And it's written like this or like this. See, it's a fraction. One over a hundred." Mr. Whyte pointed to the board. "And a decimal. Zero point zero one. So, a fraction must be less than one, too." Mr. Whyte wrote on the board.

"If you have a dime, what part of a dollar do you have? Travis?"

"Ten cents? That's a tenth of a dollar."

> $1,000 ÷ $6.15 = # of hours of work
> Decimals are less than 1 and fractions are less than 1. A penny is a hundredth of a dollar.
> $1¢ = \dfrac{1}{100} = 0.01$

"Exactly." Mr. Whyte wrote on the board some more. "It's a fraction. Ten over a hundred or one over ten—one-tenth. That's called reducing. One-tenth's decimal form is zero point one." He pointed to the board.

"So fifteen cents is fifteen over one hundred. Can someone reduce fifteen over one hundred?"

> $1,000 ÷ $6.15 = # of hours of work
> Decimals are less than 1 and fractions are less than 1. A penny is a hundredth of a dollar.
> $1¢ = \dfrac{1}{100} = 0.01$
> A dime is a tenth of a dollar.
> $10¢ = \dfrac{10}{100} = \dfrac{1}{10} = 0.10$
> So, $15¢ = \dfrac{15}{100}$

The class was silent.

"I can't hear if you all talk at once, people. Come on. Someone give it a try."

– 1.5 –

Math Speak

Decimals and fractions are less than one.

To reduce fractions to their lowest terms, you need to evenly divide the top and bottom numbers by the biggest number possible. For example, in the fraction $50/100$ both numbers can be divided evenly by 50, so $50/100$ reduces to $1/2$. In the fraction $50/75$ both numbers can be divided evenly by 25, so $50/75$ reduces to $2/3$.

Remember!
To reduce a fraction, divide the top and bottom numbers by the same number.

"Well," Sandy said slowly. "You need to find a number that you can divide by 15 and 100 evenly."

"Right. Let's start with 15. Does 15 divide into 15 evenly?"

"Yes." The class answered in unison.

"How about into 100?"

A few members of the class scribbled some numbers in their notebooks. A couple stared at the numbers on the board with lowered brows.

"No," Travis said, shaking his head.

"Nope," Phil said.

"Needs to be 5," Tom said. "5 goes into 15 and 100 evenly."

"Good for you, Tom. 15 divided by 5 equals 3, and 100 divided by 5 equals 20, so you get 3 over 20. How many nickels is that?"

$\$1{,}000 \div \$6.15 = \#$ of hours of work

Decimals are less than 1 and fractions are less than 1. A penny is a hundredth of a dollar.

$$1\cent = \frac{1}{100} = 0.01$$

A dime is a tenth of a dollar.

$$10\cent = \frac{10}{100} = \frac{1}{10} = 0.10$$

$$\text{So, } 15\cent = \frac{15}{100} = \frac{3}{20} = \$0.15$$

"Three," Jorge said.

"You got it. Three nickels, so a nickel is one-twentieth of a dollar. We'll talk about how to convert fractions to decimals another time."

Practice Problems 1-1

Reduce the fractions in the following problems to their lowest terms:

1. 25/75 = **1/3**
2. 500/750 = **2/3**
3. 60/85 = **12/17**
4. 8/100 = **2/25**
5. 54/96 = **9/16**

"Okay, let's get back to Al's car. What kind of car is it again, Al?" Mr. Whyte asked.

"It's a Mustang. It's old and kind of beat up, but the guy who's sellin' it took real good care of the engine. It needs some body work, but my cousin said he'd help me fix it."

"Good. So the car costs a thousand dollars and you're going to make $6.15 an hour. How many hours does Al need to work to make a thousand dollars?" Mr. Whyte erased everything except the top line off the whiteboard. "So we're back to $1,000 divided by $6.15."

$$\$1,000 \div \$6.15 = \text{\# of hours of work}$$

"Mr. Whyte." Jorge waved his hand in the air. "Doesn't Al need to pay taxes? Won't that mean he earns less than $6.15 an hour?"

Mr. Whyte smiled. "I was going to save that for later, Jorge. Yes, Al will need to pay taxes to Uncle Sam, so he'll earn $6.15 an hour, but he'll actually get less."

"Hey, Mr. Whyte. Why is the government called Uncle Sam?"

"I don't think I know, Olivia. Anyone think they'll have time to look it up tonight?"

"I can," Jorge said. "I can Google it."

"You guys got a computer, Jorge?" Phil asked. Phil knew that Jorge's family lost everything they owned in a fire when they lived in San Diego. It was the main reason his parents decided to move to Florida.

"No, I go to the library and use their computers. It's free."

"Good idea, Jorge. Back to the problem." Mr. Whyte tapped on the whiteboard. "We need to divide $1,000 by $6.15. Since the number that we're dividing by has a decimal in it, we need to get rid of it by moving the decimal point to the right until we have a whole number. Then we do the same thing on the other side."

$$\$1{,}000 \div \$6.15 = \text{\# of hours of work}$$

$$6.15 \overline{) 1{,}000.00}$$

2 places   2 places

Move the decimal point to the right twice.
Do it for both numbers.

"Those numbers are getting awfully big," Al said.

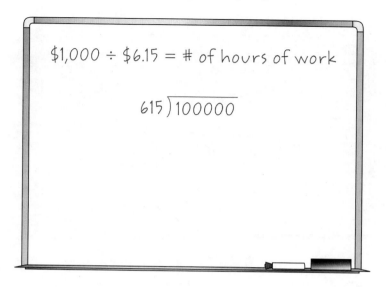

"You can do it, Al." Mr. Whyte used the eraser to cover all of the zeros. "Can 615 go into 1?"

"No."

"That's right." He moved the eraser so one zero showed. "How about 10?"

"No."

Mr. Whyte moved the eraser so two zeros showed. "How about 100?"

"No."

Mr. Whyte moved the eraser so three zeros showed. "How about 1,000?"

"Yeah, 615 goes into 1,000 once."

"You got it, Al." Mr. Whyte wrote on the board. "Subtract the 615 and you get 385. Bring the zeros down. Okay, you guys take some time and finish the problem."

 **Math Speak**

When you need to divide a whole number by a decimal number, convert the decimal number to a whole number by moving the decimal point to the right like this:

$$6.15 \overline{)1{,}000}$$

$$6.15 \overline{)1{,}000} \quad 1\rightarrow$$

$$6.15 \overline{)1{,}000} \quad 2\rightarrow$$

$$615 \overline{)1{,}000}$$

Then move the decimal point of the whole number to the right the same number of times and put zeros in the new spaces.

Remember!

$$615 \overline{)1{,}000.0} \quad \rightarrow 1$$

$$615 \overline{)1{,}000.00} \quad \rightarrow 2$$

$$615 \overline{)100{,}000}$$

Whatever you do to one number, you need to do to the other.

$$\$1{,}000 \div \$6.15 = \text{\# of hours of work}$$

$$\begin{array}{r} 1 \\ 615 \overline{)100000} \\ -615\downarrow\downarrow \\ \hline 38500 \end{array}$$

Bring the 0s down.

For a few moments, the only sound in the room was the scratch of pencils on paper, then Tom zipped open his backpack.

"What have you got there, Tom?"

"My calculator, Mr. Whyte."

"Put it away. You need to learn how to divide, not the calculator."

"Aw, come on, Mr. Whyte. In real life, I'd use a calculator."

"Think of my class as a fantasy world. No calculators."

Tom let out a long sigh and shoved his calculator into his backpack.

Practice Problems 1-2

Solve these problems to practice dividing by a decimal:

1. 25 ÷ 0.5 = **250 ÷ 5 = 50**
2. 49 ÷ 0.7 = **490 ÷ 7 = 70**
3. 56 ÷ 0.8 = **560 ÷ 8 = 70**
4. 125 ÷ 0.5 = **1,250 ÷ 5 = 250**
5. 2.5 ÷ 0.5 = **25 ÷ 5 = 5**

"Anyone want to come up to the board and give it a try? How about you, Travis?"

"I'm not done yet, Mr. Whyte."

"That's okay. Show us what you have."

Travis responded with a grimace. He slowly got out of his chair and walked to the board. His sneaker laces were untied, so there was a dull "clomp, clomp" sound as he walked. He picked up the marker and started writing.

"Tell us what you're doing," Mr. Whyte said.

"Um. I got 6 when I divided 615 into 3,850. And I got 2 when I divided the remainder. And then I got stuck."

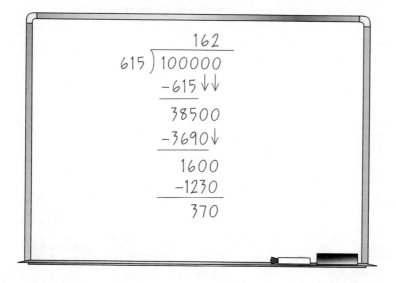

Mind Games

If you are having a problem understanding how moving the decimal point on both numbers works, let's try it on whole numbers:

$$\begin{array}{r} 5 \\ 2\overline{)10} \\ -10 \\ \hline 0 \end{array} \quad \begin{array}{r} 2.0\overline{)10.0} \\ \rightarrow 1 \rightarrow 1 \\ 5 \\ 20\overline{)100} \\ 100 \\ \hline 0 \end{array}$$

See, you get the same answer! That's because moving the decimal point of both numbers to the right once is just like multiplying both numbers by ten.

"So far so good, Travis." Travis quickly clomped back to his seat. "Anyone want to finish up?"

"You need to put a decimal point after the 162 and add a zero to the 370," Sandy said.

"Exactly." Mr. Whyte finished the problem on the board. "Okay, 162.6. Since I only have 10 left over, I'll stop there."

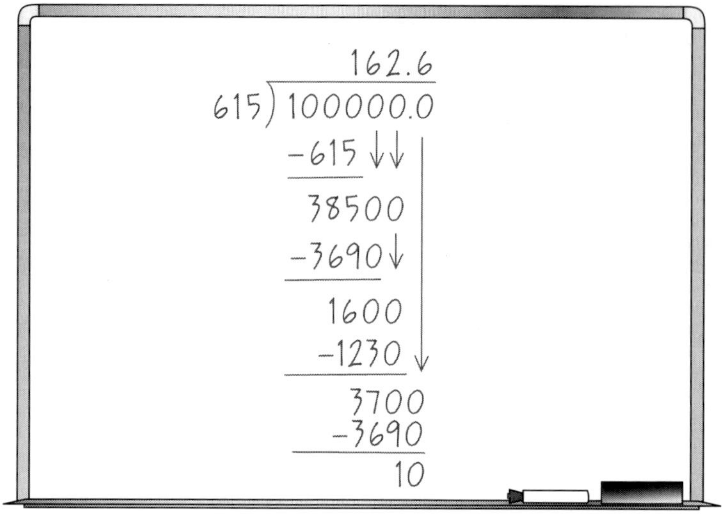

$$
\begin{array}{r}
162.6 \\
615 \overline{)100000.0} \\
-615 \\
\hline
38500 \\
-3690 \\
\hline
1600 \\
-1230 \\
\hline
3700 \\
-3690 \\
\hline
10
\end{array}
$$

"Now I need to check my answer," Mr. Whyte said. "And I do that by multiplying my answer by the number I divided by."

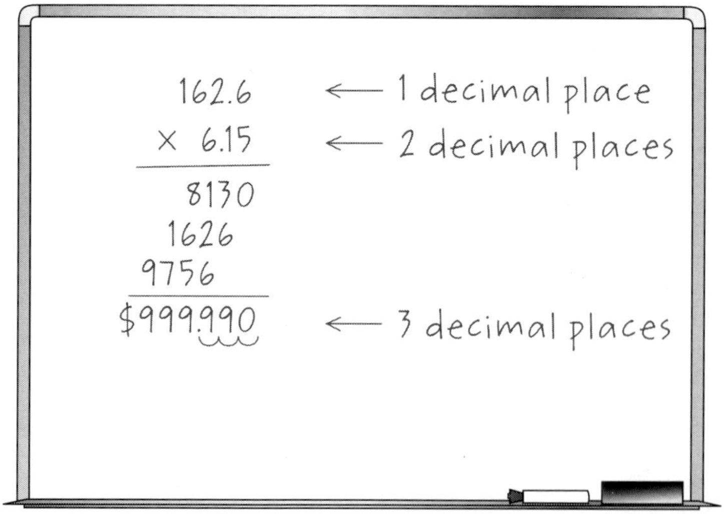

$$
\begin{array}{r}
162.6 \\
\times\ 6.15 \\
\hline
8130 \\
1626 \\
9756 \\
\hline
\$999.990
\end{array}
$$

← 1 decimal place
← 2 decimal places

← 3 decimal places

"I get $999.99. Close enough to $1,000 for me. So Al needs to work more than 162 hours to earn $1,000, but he needs to work more than that to actually get $1,000 because he needs to pay taxes."

"How much will I pay in taxes?"

"We'll save that for another day, Al. We haven't even gone over last night's homework." Just then, the bell rang. "And I guess we won't. Everyone have a good day."

The students surged towards the door. Tom and Phil hurried to catch up with Al.

"Al," Phil said. "I thought you were going to get a truck so you could help your uncle with his farm. What's with this convertible?"

"Oh, man. It's sweet. You should see it. It's red, and it's got a great sound system."

"Yeah, cool. Good thing it's a convertible. You're going to look real cool moving a cow."

Chapter 1 Review Problems

Reduce the following fractions to their lowest terms:

1. 5/100 = **1/20**
2. 20/50 = **2/5**
3. 15/45 = **1/3**
4. 56/70 = **4/5**
5. 58/100 = **29/50**

Solve these problems to practice dividing by a decimal:

6. 24 ÷ 0.6 = **240 ÷ 6 = 40**
7. 35 ÷ 0.7 = **350 ÷ 7 = 50**
8. 112 ÷ 0.8 = **1,120 ÷ 8 = 140**
9. 25 ÷ 0.25 = **2,500 ÷ 25 = 100**
10. 2.7 ÷ 0.3 = **27 ÷ 3 = 9**

 History—San Diego

 Terms

Anthropologist: A person who studies man in relation to origin, physical characteristics, environmental and social relationships, and culture.

Jorge's family is from San Diego, California. San Diego is the seventh largest city in the United States, and because of its beautiful weather, mountains, and beaches, it is a popular place to live. San Diego is the home of the world-famous San Diego Zoo, SeaWorld San Diego, and other attractions, so it is a popular vacation spot, too.

San Diego is a very old city. **Anthropologists** think that people settled in the San Diego area about 20,000 years ago. About 500 years ago, the Spanish claimed it and named it after the Spanish Catholic saint Diego de Alcalá. In Spanish, the word for male saint is San, thus the name San Diego (female saints are called Santa). Many of the old buildings in San Diego reflect this Spanish history.

When Jorge lived in San Diego, one of his favorite places to visit was the zoo. His classmates were surprised to find out that the San Diego Zoo got its start because of what is now called the Panama Canal. The canal is a manmade waterway that connects the Atlantic and Pacific Oceans and allows ships quick passage between the two oceans. When the Panama Canal was built, there were no airplanes, so freight had to be shipped by sea or by land. It took a long time to sail from the east side of the United States to the west side because ships had to go around South America.

A French company started building the canal in 1881. The work was very hard and there were so many problems, that by 1889, the company was bankrupt. The United States took over the construction in 1904 and completed the canal in 1914.

San Diego was the first port that ships coming out of the canal going to North America would reach, and the citizens of San Diego wanted to celebrate!

In 1915 and 1916, San Diego held a Panama-California Exposition, which is like a big fair. As part of the exposition, many animals from all over the world were displayed. This gave Dr. Harry Wegeforth the idea that San Diego should have a zoo, so he and some other people raised money to buy the animals after the exposition ended. That's how this world-famous zoo got started. You can read more about the San Diego Zoo at their web site (www.sandiegozoo.org).

Chapter 2
It's All About Space

Chapter 2 discusses surface area. Some students have a hard time visualizing surface area in different shapes. You can demonstrate this by cutting paper into squares and rectangles and then calculating the shape's surface area. You can even take a shape, cut it into several pieces, and then tape the shapes together to illustrate the same area taking different shapes.

This method is particularly effective to help students understand the triangle and trapezoid formulas. Take a sheet of newsprint 4 feet long and 2 feet wide, and calculate the area (8 square feet). Then cut a right triangle with a base of 1 foot from one side of the rectangle. Now you have a trapezoid and a triangle. Calculate the area of both. The sum of the areas of the triangle and trapezoid equals the area of the rectangle.

House construction has changed immensely since the birth of our nation. When early settlers first arrived, the need for housing was pressing, so houses were usually very small—sometimes as small as 10 feet by 10 feet—and considered to be temporary, although many became permanent. Small houses could be constructed quickly and were easy to heat. Whole families often lived in one or two rooms, which precluded any hope of privacy. You can look at pictures of old homes and homestead shanties online (The Plains Folk™ website is excellent).

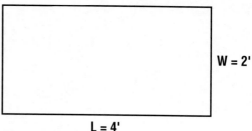

$SA = L \times W$
$= 4' \times 2'$
$= 8 \text{ ft}^2$

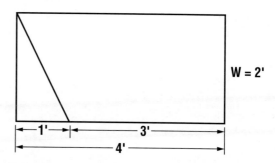

Triangle $SA = \frac{1}{2} BH$
$= \frac{1}{2}(1' \times 2')$
$= \frac{1}{2}(2)$
$= 1 \text{ ft}^2$

Trapezoid $SA = \frac{(\text{Base 1} + \text{Base 2})}{2} \times \text{Height}$
$= \frac{(3' + 4')}{2} \times 2'$
$= \frac{(7')}{2} \times 2'$
$= 7 \text{ ft}^2$

SA of Triangle and Trapezoid = $1 \text{ ft}^2 + 7 \text{ ft}^2 = 8 \text{ ft}^2$

Personalize surface area for your students. Ask them to measure their bedrooms at home and to calculate the area. Ask them if the area of their bedroom is adequate for their needs. Tell them to imagine sharing that space with one or more other people. Have them calculate the average area allotted to each occupant.

Chapter 2

It's All About Space

"Phil," Sandy raised her voice above the din of the table saw where Tom and Phil were working.

Sandy was new to Florida and started attending Lofton High School in the middle of the last school year. No one thought Sandy would last in class—everyone thought that she was too weak to handle the hard physical work, but she surprised everyone.

"What do you need, Sandy?" Tom said, pausing in his work.

"Some people are coming in to see about having a house built," Sandy said. "Mr. Whyte can't be here, so he said for Phil to talk to them."

Phil's mouth went dry and he had to swallow several times before he could get any words out. "Me? I can't talk to a homebuyer. Mr. Whyte always does that." Phil was the crew chief for the student carpenters and Mr. Whyte always said how much he appreciated that he could rely on Phil. This year the program still didn't have a buyer and everyone was getting anxious. "Can't Mr. Provost talk to them?" Mr. Provost was the assistant shop teacher.

"Mr. Whyte said for you to do it. Here, he wrote their names down for you." Sandy handed Phil a pink square of paper. "They're in the hall."

"They're here now!" Phil's eyes widened. He was sure that everyone in the shop could hear his heart pounding.

"Yeah, they're that couple near the door." Sandy tilted her head toward the shop door. Phil looked over and saw a couple in their mid-twenties. He gave Sandy a weak smile, and then slowly walked toward the door. Phil pushed his hand into his pocket to discreetly wipe the perspiration from his palm.

"Good morning." Phil tried to read the name written on the paper, but he couldn't quite get the words into focus. "Are you here to talk about building a house?"

"Yes," the man said, extending his hand. "I'm Jake Weaver, and this is my wife Ellen." Mr. Weaver was taller than his wife but not by much. He was wearing a blue uniform with a red and yellow patch on the sleeve. Phil recognized the patch; it was from a local trucking company.

"I'm Phil Tailor." He could hear his father's often repeated admonition to stand up straight, look a man right in the eye, and shake his hand firmly.

"We were expecting to see Mr. Whyte. Is he around?"

"No, sir. He was called away, so he asked me to talk to you. Let's go into the classroom. All of the stuff about the houses we build is in there."

"No offense, Phil, but we really wanted to talk to Mr. Whyte."

"If you'd rather come back..." Phil began but Mrs. Weaver interrupted him.

"Oh, Jake. Come on. We're here. We can talk to Bill."

"Um, it's Phil, Mrs. Weaver, and I really don't mind if you'd rather talk to Mr. Whyte."

"No, no. She's right. We're here. Let's talk."

"Okay." Phil pulled open the classroom door and motioned for the husband and wife to go into the room.

His stomach sank when Sandy and Tom followed them in. Phil didn't want any witnesses if he messed up. "Um, this is Sandy and Tom, Mr. and Mrs. Weaver." Phil's voice wavered and he had to clear his throat to steady it. "Do you have any ideas about what kind of house you want?"

"Not really," Mrs. Weaver said. "We didn't think we'd be able to afford a house anytime soon, but both of our parents offered to help us with the down payment, so here we are."

"Okay." Phil led the couple to a worktable in the center of the room.

"Here are some brochures about some of the houses we've built. You can look at them and see what you like."

It was cold in the classroom but Phil could feel his shirt sticking to his back. I sure hope they don't ask any hard questions, he thought.

"California Rambler," Mr. Weaver read from one of the glossy brochures. "How did you come up with that name?"

"We didn't. It's a house style. Most people call it a ranch. It's a long, narrow house with just one floor. Here's a picture of the one that the school built a few years ago. It's on North East 39th Avenue."

"It would be nice not to have to climb stairs with groceries or laundry like we do in our apartment," Mrs. Weaver said as she flipped through the floor plans.

"The guy who designed the first ranch got the idea from the adobe houses that Native Americans in California built a long time ago," said Phil. Mrs. Weaver looked bored, but Phil pressed on. "Adobe is a kind of brick, Mrs. Weaver. The Native Americans mixed mud and bits of straw together and made it into bricks. They'd let it dry in the sun, and then they'd build a house from it. It's supposed to keep it cool in the summer."

"It doesn't sound very cozy. It's like living in a root cellar. What are you looking at, Jake?"

"Cape Cod." Mr. Weaver held up a brochure with a picture of a white house with green shutters.

"Oh, that's cute. Didn't you go to Cape Cod when you were a kid?"

"Yeah, that was when I lived in New Hampshire," Mr. Weaver told Phil. "Cape Cod's in Massachusetts, and my dad loved to vacation there. He'd rent the same cottage right on the beach year after year. A lot of the houses were really small and there was always a chimney right in the middle of the roof."

Did You Know When Mrs. Weaver said an adobe home sounded like living in a root cellar, she was referring to an old-fashioned way of storing food. When America was a new country, early settlers had to grow their own vegetables and kill their own meat. They had to preserve food so they would have it for the long cold winters and mostly they dried it. These foods—like salted Codfish—needed to stay dry, so they were stored in the attic of the house. Lots of times when the family was large, the children would sleep in the attic, too!

Many years later, when times were easier, settlers had cows and chickens and they needed somewhere cool to store root vegetables like potatoes, and also milk, butter, and eggs. It is always cooler under the ground than it is above the ground, so foods that needed to stay cold were stored in the root cellar. Today, if a house has a cellar, it is called a basement and is usually clean, well lit, and lined with concrete or brick. In the 1700s and 1800s, cellars were carved out of the ground so they had earth walls and no windows. These cellars were cold, dark, and dirty, and probably bug-infested.

"That's a Cape all right. Small so they'd be easy to heat in the winter." Phil leaned against the worktable. "Do you have any idea what size house you want, Mrs. Weaver?"

"I want a small house for now, but with room to grow when we have kids."

"The Cape can be built with an unfinished second floor. The first floor is around a thousand square feet."

"I know I should know this," Mr. Weaver said. "But I was never good at math. What's this thing about square feet?"

"It's space on a surface, Mr. Weaver. Like this table." Phil waved his hand over the worktable. "Area is how much space there is on the surface of the table."

"So how would I know how much area I need in a house?" Mrs. Weaver asked.

"Come on over to the board and I'll show you." Phil picked up a marker and drew a square on the whiteboard. "This won't be to **scale**, but you'll get the idea."

"What's scale?"

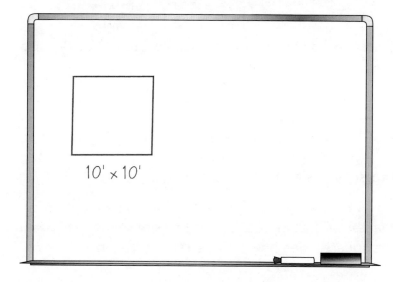

"It's this number right here," Phil said as he pointed at the numbers written on the bottom of the plan. "See, this one says 1:10. That means that every inch on the drawing is 10 feet in real life. Different plans have different scales and you need to understand the scale before you can build the house. In the old days before computers, engineers and architects drew blueprints by hand, and they used scales, which look like rulers but aren't marked in inches—they're marked by scale." Phil pointed to a couple of triangular shaped scales on the table. "Mr. Whyte's father used these scales during World War II when he was in the army. Of course, now it's all done with CAD—computer aided design."

"Interesting. So let's get back to area." Mr. Weaver pointed to the drawing on the whiteboard.

"Sure. Let's say this room is 10 feet by 10 feet—that's a square because the length and width are the same. And it's 100 square feet because you get the area by multiplying the two sides together. Now, you put a bed, dresser, bedside table, and chair in the room."

For a moment, the only sound was the squeak of the marker on the whiteboard as Phil drew some shapes in the square. "See how crowded the room looks?"

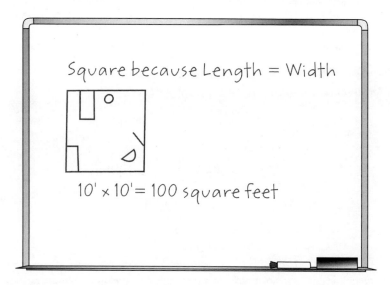

Phil used the eraser and scribbled on the board. "Now I'm going to add two feet to the width of the room. That makes the room a rectangle because the length and width aren't equal. This room is 120 square feet because its length is 10 feet and its width is 12 feet. See how much more room there is?"

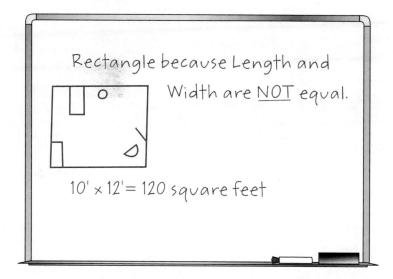

Mrs. Weaver nodded. "We'll need a bedroom that's at least that big. It looks like you have a twin bed in there, but we have a California King."

"Now you get the idea," Phil said.

Key To Understanding

Surface Area:

- Area is how much space there is on a flat surface.
- Area is measured in square units like inches, feet, and yards.
- A square has four sides that are all the same length.

L = W   Width
Length

- A rectangle has four sides too, but a rectangle's length and width are not the same.

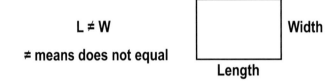

L ≠ W   Width
≠ means does not equal
Length

- Surface area for a square or a rectangle is calculated as follows:

    Length × Width = Surface Area

    L × W = SA

Since a square has equal sides, sometimes the side is labeled with a letter, like s for side, and the area is calculated as follows:

    s × s = SA or s^2 = SA

The little 2 after the s means squared. You read it as "s squared." It means to multiply s by another s.

Practice Problems 2-1

Solve these problems to practice finding the area of a rectangle or square. Draw the figure if you need help.

1. A rectangle has a length of 15 feet and a width of 10 feet. What is the area of the rectangle?

   15' × 10' = 150 square feet

2. One side of a square is 6 inches long. What is the area of the square?

   6" × 6" = 36 square inches

3. A house is 60 feet long and 30 feet wide. What is the area of the house?

   60' × 30' = 1,800 square feet

4. A Cape Cod style house has a length and width of 30 feet. What is the area of the first floor?

   30' × 30' = 900 square feet

5. A walkway has a length of 25 feet and a width of 2 feet. What is the area of the walkway?

   25' × 2' = 50 square feet

Key To Understanding

When you need to calculate the area of an irregular shape, divide the shape into common shapes.

"So that's good for a room, but what about a whole house? Like this one." Mrs. Weaver showed Phil the picture of Mr. Whyte's house. It was the first house that the class helped build. It was how the program got started. Mr. Whyte's house is an L-shaped contemporary style.

"That one is a little harder," Phil said. Phil's brow wrinkled as he looked at the floor plan and tried to figure out how to calculate the house's area. "You need to cut the outline into common shapes."

"This one is harder, Mrs. Weaver," Tom said as he stepped to the board and took the marker from Phil. "That house has a foundation that looks like this."

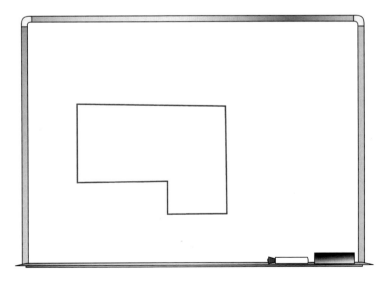

"In order to calculate the area of this house," Tom continued in the tone his father used whenever he lectured Tom. "We need to divide it into two pieces and calculate the area of each separately. Sandy, what are the **dimensions** for the top figure?"

"Looks like 50 feet by 35 feet."

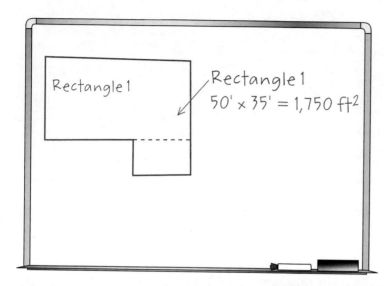

"Exactly. 50 times 35 equals 1,750 square feet," Tom said. "See, Mrs. Weaver, when I write the little two next to the feet it means square feet."

"Wow, Tom, you sound like a professor," Mr. Weaver said.

Tom turned to face the board. His neck reddened.

"So, Sandy," Mr. Weaver said with a grin. "Tell the professor the dimensions of the bottom figure." Tom's neck turned even redder.

"Looks like 15 by 12, Mr. Weaver," Sandy said.

"15 times 12 equals 180 square feet," Tom said through clenched teeth. "So the total area of the house is 1,750 square feet—which is the area of the first rectangle—plus 180 square feet—the area of the second rectangle. That's 1,930 square feet."

"Hey, Tom, you're pretty good at this math stuff," Mr. Weaver said.

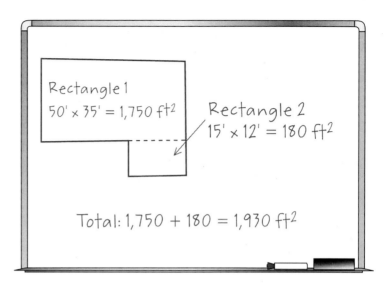

"Okay, Jake, that's enough. Stop teasing Tom," Mrs. Weaver said.

"No, I really mean it." Mr. Weaver turned to Tom. "I never paid much attention to math in school. My boss keeps telling me that I'd be a good shipping supervisor, but I need to use a lot of math—weights and volumes and stuff. I just don't get it. I wish I could do this stuff." Mr. Weaver waved a hand at the board.

"What if I know the area I need, but not the dimensions?" Mrs. Weaver asked.

"You need to know at least one dimension, Mrs. Weaver," Sandy said. "Let's say you want a room that's 300 square feet, and you know you want one wall to be 20 feet. What's next, Professor?"

"We know the formula for the area of a rectangle is length times width or L times W. So we plug in what we know. The area is 300 square feet, and the length is 20 feet." Tom drew on the board.

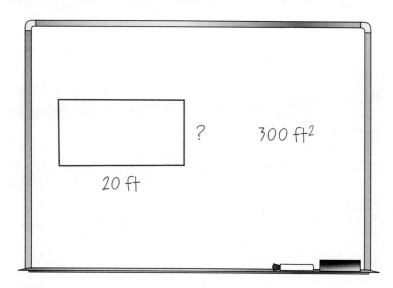

"Now the question is 20 feet times what equals 300 square feet? To find the answer, I divide both sides of the equation by 20 feet. On the right side of the equation, 20 feet divided by 20 feet equals 1, so this is cancelled out. On the left side, 300 ft² divided by 20 feet equals 15 feet, so the width is 15 feet."

$$\text{Area} = \text{Length} \times \text{Width}$$
$$A = L \times W$$
$$300 \text{ ft}^2 = 20 \text{ ft} \times W$$

$$\frac{300 \text{ ft}^2}{20 \text{ ft}} = \frac{20 \text{ ft} \times W}{20 \text{ ft}}$$

$$\frac{\overset{15 \text{ ft}}{\cancel{300 \text{ ft}^2}}}{\cancel{20 \text{ ft}}} = \frac{\overset{1}{\cancel{20 \text{ ft}}} \times W}{\cancel{20 \text{ ft}}} \qquad 15 \text{ ft} = W$$

"Tom, why did you divide? I think I would have subtracted," Mr. Weaver said.

Key To Understanding

Whatever you do to one side of an equation, you need to do to the other side to keep the equation equal.

"Dividing is the opposite of multiplying, so in this case, when you see a times symbol, you need to divide. Subtraction is the opposite of addition, so if you see a plus sign, you need to subtract. Now, I need to check my answer by multiplying 20 feet by 15 feet, and I get 300 square feet."

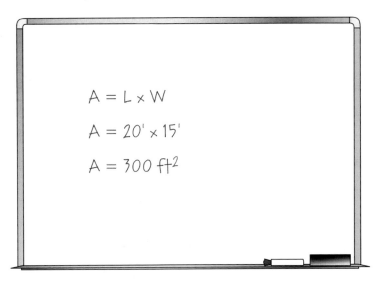

"You know what I'd really like?" Mrs. Weaver interrupted. "Do you remember the window seat at my grandmother's house?" She turned to Sandy. "It had three windows. I don't remember what you call it."

"It's a bay window," Sandy said.

"A what?" Phil asked.

"Bay window. Here, there's a picture of one in this home decorating magazine." Sandy flipped through a glossy magazine and held up a picture.

"Yes. That's it," Mrs. Weaver exclaimed. "Except my grandmother's had cushions and pillows on the seat and these long blue drapes that looped back. I could unloop the drapes and close myself in. I loved curling up and reading in that window seat."

"What kind of shape is that, Sandy?" Tom asked.

Sandy smiled. "Why Tom, that's a trapezoid. You know that." Sandy picked up the marker and drew a four-sided figure on the whiteboard.

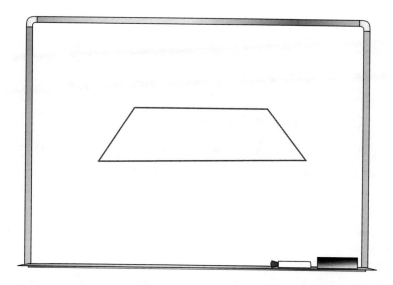

"A trapezoid has four sides with two of the sides being parallel. Parallel means that the lines are always the same distance apart, even if they go on forever."

"Sandy, can you figure out the area of a trapezoid?" Mr. Weaver asked.

"You bet. The parallel lines are called base 1 and base 2. So let's say that the window side is base 1 and it's four feet. And the front edge of the seat is base 2 and is ten feet. That makes it nice and big. Now, for the width of the seat." Sandy studied the picture in the magazine. "This one looks a little narrow to me, so let's make ours two feet. That means the height of the trapezoid is two feet. This is the formula to find the area of a trapezoid."

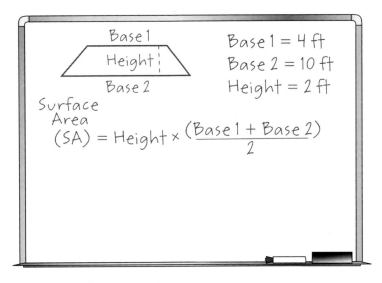

"Wait a minute, Sandy. Why do you add the bases together and then divide by two?" Mr. Weaver pointed to the formula that Sandy had written on the board.

"That's how you **average** the numbers, Mr. Weaver. You add two numbers together and then divide the answer by two. When you add three numbers together and then divide the answer by three that's averaging, too. Now let's plug in the numbers and do the equation."

Sandy scribbled on the board. "The surface area of the bay window seat is fourteen square feet."

"You know what, Sandy? That averaging doesn't seem right to me. How do you know it's right?" Mr. Weaver asked.

[Whiteboard:]

Base 1 = 4 ft
Base 2 = 10 ft
Height = 2 ft

$$SA = Height \times \frac{(Base\ 1 + Base\ 2)}{2}$$

$$SA = 2\left(\frac{4 + 10}{2}\right)$$

$$SA = 2\left(\frac{14}{2}\right)$$

$$SA = 2(7) \qquad SA = 14\ ft^2$$

Key To Understanding

Trapezoids

A trapezoid is a four-sided shape. It has two sides that are parallel and two that are not. These figures are examples of trapezoids:

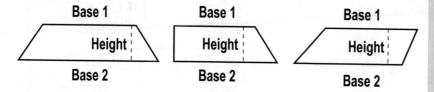

The parallel sides are called Base 1 and Base 2. To figure out the surface area of a trapezoid, you need to average the lengths of the bases and multiply the base average by the height:

$$\left(\frac{Base\ 1 + Base\ 2}{2}\right) \times Height$$

Average means to find the midpoint between numbers. (Average is also called mean.) To find the average of two or more numbers, add all of the numbers and then divide by the quantity of numbers you added. To average these four numbers: 3, 7, 15, and 19, add the numbers: 3 + 7 + 15 + 19 = 44 and then divide 44 by 4, which is 11 (44 ÷ 4 = 11).

Practice Problems 2-2

Solve these problems to practice finding the area of a rectangle or square. Draw the figure if you need help.

1. A trapezoid has dimensions of base 1 = 2 inches, base 2 = 4 inches, and a height of 6 inches. What is the area of the trapezoid?

   [(2" + 4") ÷ 2] × 6" = 18 square inches

2. What is the area of the following figure?

   [(15' + 21') ÷ 2] × 12' = 216 square feet

   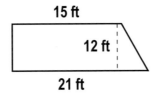

3. Mrs. Weaver wants to have a bay window in her living room, so the living room outline looks like this:

   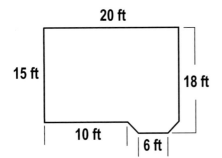
   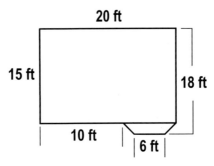

   Divide the living room outline into a rectangle and a trapezoid and label the dimensions of the two shapes.

4. Calculate the areas of the rectangle and the trapezoid.

   Rectangle = 20' × 15' = 300 square feet

   Trapezoid = [(6' + 10') ÷ 2] × 3' = 24 square feet

5. What is the total area of the living room?

   300' + 24' = 324 square feet

"I can prove it using triangles and rectangles. Here's the 10 by 12 rectangle Phil used earlier." Sandy scribbled on the board. "Remember how we calculated the area—10 times 12? If I cut it in half diagonally, I get two triangles. Triangles are three-sided figures. What's the area of each triangle?"

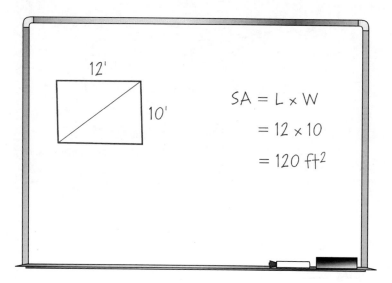

"It has to be half, so that's 60 square feet," Mr. Weaver answered.

"You got it, so you calculate the area of a triangle by multiplying the length by the width and dividing by two."

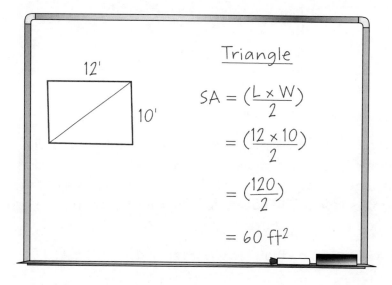

Key To Understanding

Triangles

A triangle is a three-sided figure that has three angles.

The angles in the triangle may vary, but the sum of the angles is always 180°.

The right triangle is one of the most frequently used in construction. A right triangle must have one 90° angle, so the sum of the other two angles must equal 90°. A right triangle is created when a square or rectangle is divided in half diagonally so that you have two identical triangles, each with half the surface area of the square or rectangle.

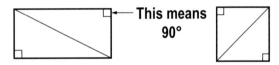

A triangle's length is called the base, which is abbreviated B. Its width is called the height, which is abbreviated H, so the formula to calculate the surface area of a triangle is 1/2BH. This formula is used to calculate the surface area of all triangles.

"Okay, but what does this have to do with a trapezoid?"

Sandy smiled at Mr. Weaver. "When I redraw the diagram so that the triangles are on each side of the rectangle, I get a trapezoid. And I already know the area of the whole shape is 240 square feet, because each triangle is 60 square feet and the rectangle is 120 square feet."

"I'm with you on that," Mr. Weaver said.

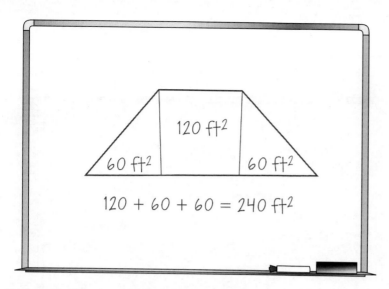

"I'll call the top side base 1, which I know is 12 feet 'cause that's the length of the rectangle. Base 2 is the bottom side and it has to be 36 feet—12 feet from the rectangle and 12 feet from each of the triangles."

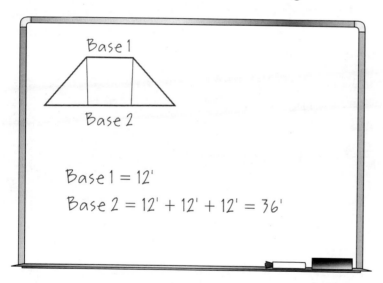

Mrs. Weaver pointed to the board and counted. "12 feet from the rectangle, and 12 feet from the triangle, and 12 feet from the other triangle. So that means base 2 is 36 feet." She nodded. "Yes, I agree with that."

"Good," Sandy continued. "The height is 10 feet because that was the width of the rectangle. Now, using

 **Mind Games**

A right angle equals 90°. Another way to say right angle mathematically is perpendicular. When a line is perpendicular to another line, it means they meet at a 90° angle.

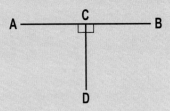

Line CD is perpendicular to AB.

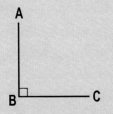

Line AB is perpendicular to BC.

the trapezoid surface area formula, we need to plug in the values for bases 1 and 2, and the height. And then do the math, and voila! 240 square feet."

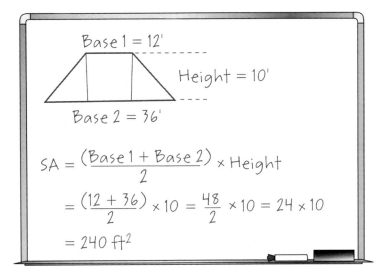

Mr. Weaver pointed at the figures on the board. "12…36…48…240. Yeah, I can see how you get 240 square feet, but this averaging still has me baffled."

"Mr. Whyte has us remember that by using a rectangle and averaging its lengths." Sandy scribbled on the board. "See."

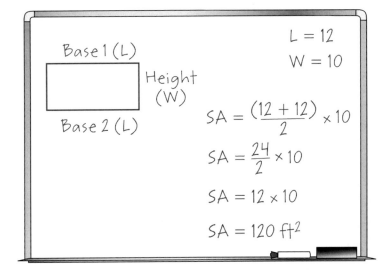

Mr. Weaver smiled. "Yeah, I do. You guys are good. I wish I could be in Mr. Whyte's class."

"You know, Mr. Weaver," Tom said. "Mr. Whyte teaches adult math every autumn at the community education center. It starts next week."

"Jake, we need to get going," Mrs. Weaver interrupted as her husband was about to speak. "Phil, tell Mr. Whyte we'll call him. Okay?"

"Yes, ma'am."

"Tom, can you show us how to get to the front door? We got all turned around when we came in. And you can tell us about this math class. I always wanted to be a nurse, but I thought the math was too hard. But I caught most of what you guys showed us. Maybe there's hope for me."

"Sure, Mrs. Weaver, I'll be happy to show you out," Tom said as he followed the couple to the door.

Chapter 2 Review Problems

1. The outline of the first and second floors of a Cape Cod style house are shown below. Calculate the total area for the house.

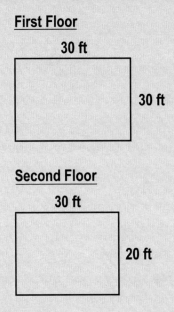

First Floor
30' × 30' = 900 square feet
or 30² = 900 square feet

Second Floor
30' × 20' = 600 square feet

Total
900 square feet + 600 square feet = 1,500 square feet

2. The outline for a ranch style house is a simple rectangle that is 60 feet long and 30 feet wide. What is its area?

60' × 30' = 1,800 square feet

3. The Weavers really liked the design of Mr. Whyte's house, but Mrs. Weaver wanted to add a large bay window. The outline of the house is shown below. Describe in three or more sentences how you would calculate the surface area of the house. (Hint: Start with the sentences: "To find the area of the house, I need to divide the house into common shapes. They are ...")

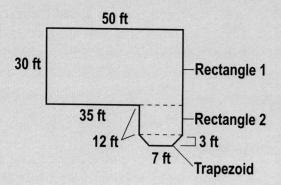

Answers will vary. Stress that the first sentence is a topic sentence and explains what needs to be done (find the surface area of the house). The last sentence is the conclusion sentence and needs to include the topic. The sentences between need to move logically from the topic to the conclusion. For example:

To find the area of the house, I need to divide the house into common shapes. They are two rectangles and a trapezoid. I need to figure out the area of each shape and then add the three results to get the total surface area of the house.

4. Divide the house into the shapes you described in Problem 3 and label all dimensions. Calculate the area of each of the shapes.

   Rectangle 1 = 30' × 50' = 1,500 square feet

   Rectangle 2 = 12' × (50' − 35') = 180 square feet

   Trapezoid   = [7' + (50' − 35') ÷ 2] × 3 = 33 square feet

5. Calculate the floor space (area) of the entire house.

   1,500' + 180' + 33 = 1,713 square feet

 **Terms**

Average: The middle point between two numbers or the mean of two or more numbers. It is calculated by adding all the numbers together, and then dividing the sum by the quantity of numbers added. For example, the average (or mean) of 3, 7, and 11 is 7 (3 + 7 + 11= 21; 21 ÷ 3 = 7).

Dimensions: The measurements of a shape or object. The dimensions of a rectangle are the measurements of its length and width. The dimensions of a triangle are the measurements of all three sides. A square has two dimensions, but you only need one because both dimensions are the same.

Scale: A proportion between two sets of dimensions—like the dimensions on a drawing and the actual dimensions of the object.

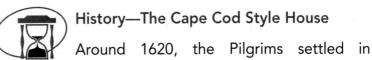 History—The Cape Cod Style House

Around 1620, the Pilgrims settled in Plymouth, which is on Cape Cod. It would be nice to say that they lived in Cape Cod style houses, but such houses were too luxurious for that time. These houses were built much later, when more people came over to Cape Cod from England. They wanted houses like the ones in England, so a Cape Cod house looks a lot like an English cottage.

Careers in Construction—CAD Operators

Computer aided design (CAD) systems use computer programs to generate drawings for construction. CAD systems are fast, easy to use, and accurate. They make it easy to make changes to a drawing. CAD operators need to be good with computers, but that is only part of it. Operators need to be able to visualize a design in their head so that they can put it on paper, and they need to be good at math. Some CAD operators do not have any special education—they learned the CAD system by using it. Others may have taken one or more classes to learn about CAD systems. Some even have a degree. If you want to learn more about education for CAD operators, contact your local community college.

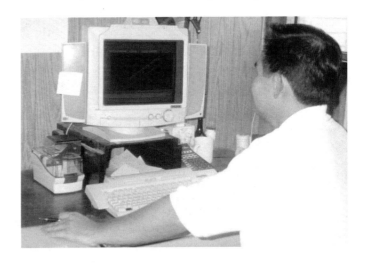

Chapter 3
Where Do You Live?

This chapter discusses cost per square foot, which is an average. Students need to understand long division and decimals. This chapter also introduces powers of ten.

Relate averages to a student's class average. Give the students a list of test scores and help them find the overall average. In advanced classes, you can introduce weighted averages.

Some students need to see something to truly understand it. You can demonstrate averages using several measuring cups (or beakers) that contain different levels of water. Show the class how to total the levels and then find the average. Do this demonstration several times with different levels of water. End the demonstration by showing students how one very low level of water pulls down the average by emptying a cup (some students get this when using test scores).

Real estate costs vary dramatically by region. In some areas like Boston, New York City, Los Angeles, San Francisco, and so on, real estate prices can vary by neighborhood. This is a good time to discuss some of the reasons for these variations. Some less obvious reasons are population density, grandfathered building codes, historic areas, desirable schools, and availability of public transportation.

It is important to explain to the students that cost per square foot is simply a rule of thumb. Actually, the cost per square foot to build the same house can vary a great deal. Energy efficient windows, high-efficiency heating/cooling systems, granite countertops, expensive landscaping, ceramic tile floors, and architectural touches like arched doorways can add significantly to the initial cost of a home.

Land costs also vary widely. A building lot in a nice residential subdivision can cost $100,000 or more, while a building site out in the country might cost a tenth of that amount for a plot ten times the size. So, if the price of the lot is included in the calculation, the cost per square foot for the same house in two locations would be very different. Much of it depends on who is doing the calculation, however. In order to determine the true cost of building a house, a builder would subtract the cost of the land. But real estate agents and homebuyers often use the total cost as a means of comparing homes that are up for sale.

You can apply this information to your hometown. Use the classified section of the newspaper to show the students the variations in real estate prices. Pick out some homes that the students like and calculate the cost per square foot. Look for the price of vacant lots and compare their prices to those of acreage outside a populated area.

Chapter 3

Where Do You Live?

"Hi, Travis," Phil said. "Did you get your homework done?"

"Yeah, it was pretty easy."

"Easy? Mine was hard. What city did you get?"

Mr. Whyte had assigned homework on Friday. Each student was assigned a city someplace in the United States and had to search online real estate sites for three houses they liked. Then they had to figure out the housing cost per square foot.

"I got Kearney, Nebraska," Travis said with a laugh. "I had to look up Nebraska in my Dad's road atlas."

"I got a place in Massachusetts called the Back Bay. It's part of Boston, and houses are really expensive there." Phil turned to Sandy. "Hi, Sandy. We missed you Friday. Were you sick?"

Sandy hugged her books close to her body as if she was cold and forced a smile at the boys. She shook her head. "No...um...my mom needed me at home."

Sandy joined the group of students hurrying into the classroom, took her seat, and then busied herself with her notebook. Phil and Travis glanced at each other and Phil shrugged. Sometimes Sandy was friendly and other times she wasn't.

Mr. Whyte swung into the room. "How did you guys like the homework assignment?"

"Mr. Whyte, some of those numbers were awfully big! There must be an easier way to divide big numbers," Al said.

"There is. It's called a calculator." Tom grinned at Mr. Whyte.

"Persistence, Tom, that's what I like." Mr. Whyte punched the air. "Al, you don't know it, but you just introduced today's lesson—we're going to talk about powers of ten. But let's get the homework started first. What's cost per square foot mean to us? Tom?"

"It's an average. You spread out the cost of the house evenly over the whole house."

"Does it mean each square foot costs the same amount?" asked Mr. Whyte.

"No, sir. It's an average. Just like when you calculate the class average on a test. Some people get high grades and some low, but the class average is one number, it's what everyone would get if the scores were spread out evenly."

"Good. What are the most expensive rooms in a house?" Mr. Whyte asked.

"The kitchen and bathroom!" The class answer in unison. It was one of Mr. Whyte's favorite questions.

"Who says you guys don't listen? You calculate average cost per square foot by dividing the total cost by the area, which can be written three ways. Since it's money, I asked you to carry the division out to the second decimal place."

$$\text{Average Cost} = \text{Area} \overline{)\text{Cost}}$$
$$(\text{ft}^2) \quad\quad (\text{ft}^2)$$

$$= \frac{\text{Cost}}{\text{Area (ft}^2)}$$

$$= \text{Cost} \div \text{Area (ft}^2)$$

"Okay, guys, homework. Travis, Olivia, and Al. Up to the board." The students knew the drill. Mr. Whyte often had them write their homework assignments on the board—it was his way of making sure they at least tried to complete the assignment. "Okay, Olivia. You have the floor."

"I got Dayton, Ohio." Olivia began, pointing to her neat writing on the board. "I found a nice two-story house with three bedrooms and two bathrooms for $137,000. It has an area of 1,488 square feet. I divided $137,000 by 1,488 to get $92.07 per square foot."

```
                    $92.069 per ft²
          1,488 ft² ) $137,000.000
                     13392
                      3080
                      2976
                     10400
                      8928
                     14720
                     13392
                      1328
```

Dayton, Ohio

Rounded to $92.07 per ft²

"Good, Olivia. Somebody check her work." Mr. Whyte made his students show all of the work to solve a problem, and sometimes he made them check the work on the board, but usually he had someone check homework answers on a calculator.

"Looks good, Mr. Whyte," Phil said.

"Good job, Olivia. Travis?"

"I got Kearney, Nebraska. I found a 1,008 square foot house for $39,900, so I figured the average cost to be $39.58 per square foot."

```
                    $39.583 per ft²
         1,008 ft² )$39,900.00
                    3024
                    ─────      Kearney, Nebraska
                    9660
                    9072       Rounded to
                    ─────
                    5880         $39.58
                    5040
                    ─────       per ft²
                    8400
                    8064
                    ─────
                    3360
                    3024
                    ─────
                     336
```

"Whoa, Travis, that's awfully cheap. Is the place a dump?" Phil asked.

"Naw, I thought it was nice. See." Travis held up a picture. "It's a little small, but I liked it. I picked out three other houses and none of them were as high as Olivia's house."

"You guys are bringing up a good point," Mr. Whyte said. "Costs vary by region. After we finish the homework, I want to talk about that, so start thinking about reasons why there is such a big cost difference. Al, you're on."

"Well, I got New Yawk, New Yawk," Al said with a grin. "I got cousins from there, but they don't really talk like that. Anyway, I picked out a nice big place in Manhattan—that's an island—it's 3,500 square feet and only costs about $5 million dollars—$4,970,000 to be exact."

"That must be some house," Olivia said.

"Hey, I'm not sure it's a house. It could be an apartment, or … what do you call it, Mr. Whyte?"

"In New York? Maybe a **co-op**. What did you get for the average cost per square foot?"

"$1,420.00."

A few students let out whistles.

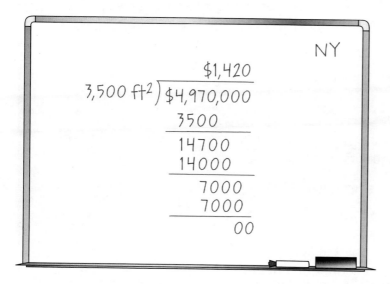

"Thanks, guys," Mr. Whyte said to the three students standing near the board. "You may sit down now."

Practice Problems 3-1

1. Jorge found a house he liked in San Diego. It's 1,918 square feet and costs $829,000. What is the average cost per square foot?

   $829,000 ÷ 1,918 = $432.221 per square foot

2. Olivia found another house she liked in Dayton. It costs $74,900 and is 1,092 square feet. What is the average cost per square foot?

   $74,900 ÷ 1,092 = $68.589 per square foot

3. Another student liked a house in Hyannis, Massachusetts. It's really small at 485 square feet and it costs $218,900. How much does it cost per square foot?

   $218,900 ÷ 485 = $451.340 per square foot

4. Travis also liked a bigger house in Kearney, Nebraska. It's 2,052 square feet and costs $99,900. What is the cost per square foot?

   $99,900 ÷ 2,052 = $48.684 per square foot

5. Mr. Whyte liked a small apartment in Brooklyn, New York. It costs $659,000 and is 576 square feet. This apartment comes with a deeded parking place! What is the cost per square foot?

   $659,000 ÷ 576 = $1,144.097 per square foot

"Al's dream house costs a lot," Mr. Whyte said. "This is a good time to talk about exponents. The numbers in Al's problem are pretty big, so I want to make them smaller. You all know about squaring numbers." Mr. Whyte scribbled on the board.

$$2^4 = 2 \times 2 \times 2 \times 2 = 16$$
$$2^3 = 2 \times 2 \times 2 = 8$$
$$2^2 = 2 \times 2 = 4$$
$$2^1 = 2$$
$$2^0 = 1$$

"Anybody have a problem with this?" Mr. Whyte asked, pointing to the board. Most students shook their heads no.

"Now bear with me here." Mr. Whyte continued writing on the board.

$$10^4 = 10 \times 10 \times 10 \times 10 = 10,000$$
$$10^3 = 10 \times 10 \times 10 = 1,000$$
$$10^2 = 10 \times 10 = 100$$
$$10^1 = 10$$
$$10^0 = 1$$

Math Speak

In the expression 2^2, the exponent is the little number 2. The base is the big number 2. So in the expressions 3^1, 4^5, and 5^6, the little 1, 5 and 6 are the exponents, and the big 3, 4, and 5 are the bases.

When you see an expression like 4^5, you say 4 raised to the 5th power. When you see this expression, you need to multiply the base number by itself. Write down the base the number of times shown in the little number. In this case:

4 4 4 4 4

Then put a multiplication sign between the numbers:

4 × 4 × 4 × 4 × 4

Now multiply the numbers:

4 × 4 × 4 × 4 × 4

4 × 4 = 16

16 × 4 = 64

64 × 4 = 256

256 × 4 = 1,024

So that means:

4^5 = 4 × 4 × 4 × 4 × 4 = 1,024

"Can everyone agree with this?" Mr. Whyte tapped the board with his marker.

A murmur of yeses and sures rippled through the room.

"This is called base 10 because the base number is 10 and these are the powers of 10. Powers of 10 is a handy thing to use when you have a large number with lots of zeros on the end." He erased the board and wrote 1,000,000.

"The way you convert a large number to a power of ten is to move the decimal point to the left until you reach the end of the trailing zeros."

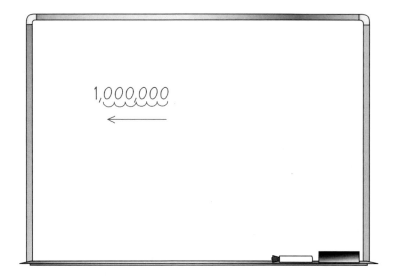

"Then you count the number of places that you moved the decimal point, and write that number as the exponent of the 10."

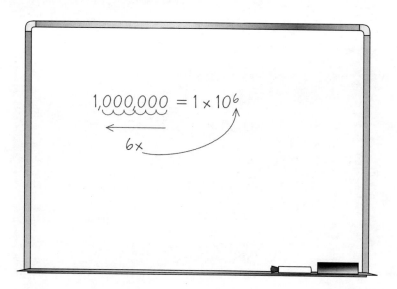

"You don't have to get rid of all the zeros. You move the decimal point as many or as few times as you want. This is called scientific notation." Mr. Whyte scribbled on the board.

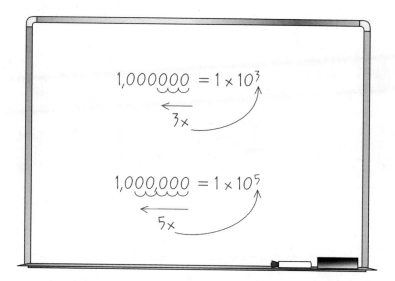

Mind Games

Lots of students have trouble with the first power (2^1) and zero power (2^0). When you see the first power, it means write the base number only once, but when you see the zero power, it means divide the base number by itself.

If you still have problems remembering this, perform the following exercises:

2^4 = **16**

$16 \div 2$ = **8**

2^3 = **8**

$8 \div 2$ = **4**

2^2 = **4**

$4 \div 2$ = **2**

2^1 = **2**

$2 \div 2$ = **1**

2^0 = **1**

Practice Problems 3-2

Convert the following numbers as directed.

1. $1 = 1 \times 10^{\underline{0}}$
2. $10,000 = 10 \times 10^{\underline{3}}$
3. $100,000 = \underline{\mathbf{100}} \times 10^3$
4. $1,000 = 1 \times 10^{\underline{3}}$
5. $1,000,000 = 10 \times 10^{\underline{5}}$
6. $10,000 = \underline{\mathbf{1,000}} \times 10^1$
7. $10,000 = 1 \times 10^{\underline{4}}$
8. $100,000 = \underline{\mathbf{1}} \times 10^5$
9. $1,000 = 10 \times 10^2$
10. $1,000,000 = 100 \times 10^{\underline{4}}$

Mr. Whyte continued. "Now let's convert Al's numbers to powers of 10."

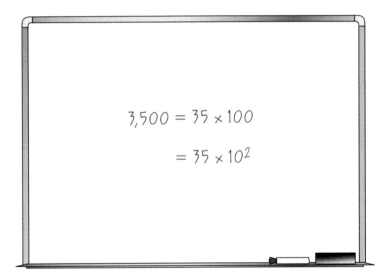

$$3,500 = 35 \times 100$$
$$= 35 \times 10^2$$

"3,500 is equal to 35 times 100, so I just substitute 10^2 for the 100 and now I have 35 times 10 to the second power. Now the other number."

He looked back at the students to see if anyone was lost. "Are we still okay? Everyone get it? All I'm doing is substituting a power of ten for a number."

$$4{,}970{,}000 \quad \begin{aligned} &= 497 \times 10{,}000 \\ &= 497 \times 10^4 \end{aligned}$$

OR
$$\begin{aligned} &= 4{,}970 \times 1{,}000 \\ &= 4{,}970 \times 10^3 \end{aligned}$$

OR
$$\begin{aligned} &= 49{,}700 \times 100 \\ &= 49{,}700 \times 10^2 \end{aligned}$$

OR
$$= 497{,}000 \times 10$$

"Wait a minute, Mr. Whyte. Let me think about it for a second," Al said. "10,000 equals 10^4, so 497 times 10^4 equals 4,970,000…yeah, got it."

"Let's press on, but anyone who doesn't get it, see me after class. Let's plug some powers of ten into the problem. You all know how to divide whole numbers, so I want you to ignore the 35 and 497 for the moment and concentrate on the powers of ten."

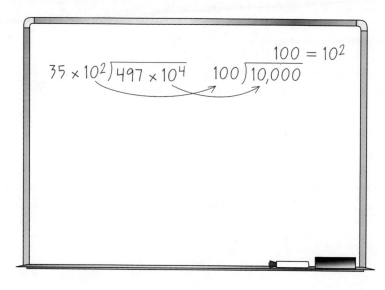

Math Speak

In math, division can be written in different ways. For example, 12 divided by 3 can be written:

$12 \div 3 \qquad 3\overline{)12} \qquad \frac{12}{3}$

The number to be divided, in this case 12, is called the dividend.

The number that you are dividing by, in this case 3, is called the divisor.

The answer to a division problem is called a quotient.

Mind Games

Some people have a hard time remembering which numbers are the dividends, divisors, and quotients. You can remember it this way: when you write the problem 12 ÷ 3 = 4 or 12/3 = 4, the words are in alphabetical order (dividend, divisor, and quotient).

"The 10^2 is the same as 100 and the 10^4 is the 10,000, so 10,000 divided by 100 equals 100, and 100 equals 10^2." Mr. Whyte pointed to the board. "Questions?"

Phil stared at the numbers on the board with lowered brows. He blew out a sigh and said, "I see what you're doing, Mr. Whyte, but I don't see where this is going."

"Patience, Phil, patience. Let's add a couple of more examples."

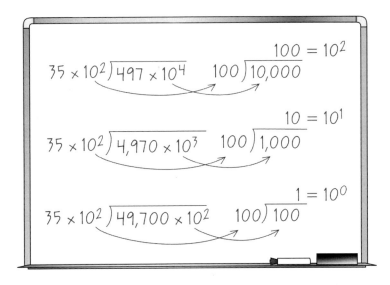

"Anyone see a pattern here? Jorge?"

Jorge leaned back in his chair, crossed his arms, and studied the board with narrowed eyes. He slowly nodded his head. "You're subtracting the exponent in the divisor from the exponent in the dividend to get the answer."

"Bingo, Jorge."

"Let's look at that again. I take the 2 from the divisor and subtract it from the 4 in the dividend and get 2, so in my quotient I write 10 to the 2nd power."

Math Speak

When you need to divide with powers of 10, subtract the exponent in the divisor from the exponent in the dividend to get the power of 10 for the quotient. Notice the base number does not get divided.

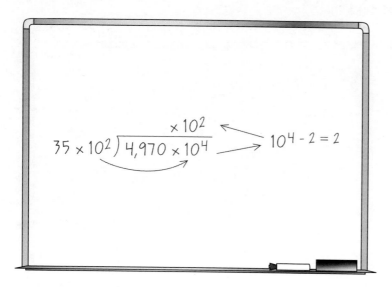

Practice Problems 3-3

1. $1 \times 10^4 \div 1 \times 10^4 =$
   $1 \times 10^{\underline{0}}$

2. $1 \times 10^8 \div 1 \times 10^6 =$
   $1 \times 10^{\underline{2}}$

3. $1 \times 10^{15} \div 1 \times 10^{12} =$
   $1 \times 10^{\underline{3}}$

4. $1 \times 10^9 \div 1 \times 10^8 =$
   $1 \times 10^{\underline{1}}$

5. $1 \times 10^2 \div 1 \times 10^0 =$
   $1 \times 10^{\underline{2}}$

"To complete the problem, you divide the numbers—in this case, 4,970 divided by 35, to get 142 times 10 to the 2nd power. So the answer will be 142×10^2, or 14,200."

$$\begin{array}{r} 142 \times 10^2 \\ 35 \times 10^2 \overline{\smash{)}\, 4{,}970 \times 10^4} \\ \underline{35} \\ 147 \\ \underline{140} \\ 70 \\ \underline{70} \\ 0 \end{array}$$

"Cool. That sure makes it easier to divide big numbers," Al said.

"Mr. Whyte," Jorge said. "This cost per square foot doesn't seem right to me. When you buy a house, you pay for the house and the land together."

"Jorge, you have the mind of an accountant." Mr. Whyte smiled. "You're right. I assigned the homework to illustrate average cost. Real estate agents use the method we used—land and buildings together. But builders are only interested in the cost to build the house. Builders calculate the average cost per square foot of a house without the cost of the land."

"How can you buy a house without buying the land, Mr. Whyte?" Olivia asked.

Practice Problems 3-4

In the following problems, convert the cost and area to a power of 10 and then calculate the average cost per square foot.

1. Jorge found another house he liked in San Diego. It's 1,000 square feet and costs $5,500,000. What is the average cost per square foot?

   $5,500 × 10^3 ÷ 10^3 = $5,500 per ft^2

2. Phil found a small house he liked in the Hyde Park section of Boston, Massachusetts. It costs $198,000 and is 600 square feet. What is the average cost per square foot?

   $1,980 × 10^2 ÷ 6 × 10^2 = $330 per ft^2

3. Another student found a tiny apartment in Brooklyn, New York. He didn't really like it, but since it cost so little compared to other real estate in that area, he was curious about how much it cost per square foot. It's 300 square feet and costs $325,000. How much does it cost per square foot?

   $3,250 × 10^2 ÷ 3 × 10^2 = $1,083.333 per ft^2

4. Al has expensive tastes. Another house he picked out costs $1,100,000 and is 2,200 square feet. What is the cost per square foot?

   $1,100 × 10^3 ÷ 2.2 × 10^3 = $500 per ft^2

5. Mr. Whyte found a house in Gainesville, Florida, not far from Lofton High School. It costs $109,900 and is 1,100 square feet. What is the cost per square foot?

   $1,099 × 10^2 ÷ 11 × 10^2 = $99.909 per ft^2

"Sometimes developments are built on leased land. Instead of the homebuyer purchasing the land, the landowner attaches a lease—for say, 100 years—to the house. And the homebuyer pays the landowner a monthly or yearly rental fee." Mr. Whyte said. "What Jorge is saying is that when you buy a house you get two pieces of property—a house and the land it's built on. To calculate the cost per square foot of the house, you need to subtract the value of the land to get the cost of the house and then divide the area by the house cost."

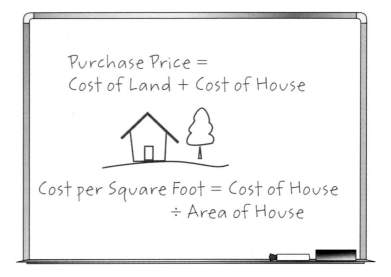

"Al, you're looking perplexed."

Al leaned forward in his chair. "I was just thinking about my cousins in New York. New York is really crowded, not like here. If there are a lot of people in an area, I'll bet that makes land prices high."

"You guys don't need me. You're figuring all of this stuff out for yourself. What about the hurricanes in south Florida? How could that affect housing costs?"

"Makes building materials hard to get so the price goes up," Travis said.

"Good, Travis, simple economics—supply and demand. High demand, high prices," Mr. Whyte said.

"Building codes," Sandy mumbled. Mr. Whyte could see she was upset about something.

"That's right. When an area gets a lot of bad weather, local governments require strong buildings. All that extra strength costs money. Class is almost over. Jorge, did you find out anything about **Uncle Sam**?"

Jorge sat up straight. "Yes, sir." Jorge told his classmates about Uncle Sam until the bell rang.

Phil worked his way through the crowd of students at the door to the hall window. He rested his books on the sill and bent to tie the lace of his sneaker. The noise thinned as his classmates made their way down the hall and Phil could hear Mr. Whyte talking to someone in the classroom. Phil didn't mean to eavesdrop, but he found himself straining to hear what was being said.

"Are you doing okay?"

There was some indistinct murmuring as the student answered.

"Do you need to go home?"

Mr. Whyte's voice was getting louder as he walked toward the door. Phil grabbed his books and sprinted down the hallway. As he turned the corner, he caught a glimpse of Sandy coming through the classroom door. What's that about?, he wondered.

Chapter 3 Review Problems

Convert the following numbers as directed.

1. $1{,}000{,}000 = 1{,}000 \times 10^3$

2. $100{,}000 = \underline{\mathbf{100}} \times 10^3$

3. $1{,}000 = 1{,}000 \times 10^{\underline{0}}$

4. $1{,}000{,}000 = 10 \times 10^{\underline{5}}$

5. $(256 \times 10^4) \div (8 \times 10^4) = \underline{\mathbf{32}} \times 10^{\underline{0}}$

6. $(125 \times 10^8) \div (5 \times 10^6) = \underline{\mathbf{25}} \times 10^{\underline{2}}$

7. $(999 \times 10^{15}) \div (333 \times 10^{12}) = \underline{\mathbf{3}} \times 10^{\underline{3}}$

Using the formula Average Population = Total Population ÷ Total Area, calculate the average number of people per square mile for the zip code areas in Questions 8 and 9.

8. Al lives in zip code 32609. It has a population of 18,208 and a land area of 101.3 square miles. What is the average population per square mile rounded to the nearest whole number?

   18,208 people ÷ 101.3 square miles = 179.743 = 180 people per square mile

9. One of the houses Al liked was located in zip code 10035 (NY). It has a population of 32,702 and an area of 1.4 square miles. What is the average population per square mile rounded to the nearest whole number?

   32,702 people ÷ 1.4 square miles = 23,358.571 = 23,359 people per square mile

In Questions 10 and 11, calculate the average number of people per household from the information given.

10. In zip code 32609, the population is 18,208 and there are 7,449 households (houses, apartments, condominiums, etc.).

    18,208 people ÷ 7,449 households = 2.444 people per household

11. In zip code 10035, the population is 32,702 and there are 12,054 households (houses, apartments, condominiums, etc.).

    32,702 people ÷ 12,054 households = 2.713 people per household

12. Compare the populations in zip codes 32609 and 10035. Which area is more crowded? Describe what you think the neighborhoods must look like based on the population.

    Zip code 10035 is clearly far more crowded than zip code 32609. Zip code 32609 is likely to have single-family homes, with large yards, while zip code 10035 will probably be multi-family dwellings—like apartment buildings.

 Terms

Co-op: Many large cities have cooperative apartments, called co-ops. In co-ops, a corporation owns the apartment building and the owners all buy shares of the corporation. All owners pay a monthly fee, which includes the mortgage payment and a maintenance fee for the upkeep of the building.

Uncle Sam: Uncle Sam was named for Sam Wilson. Mr. Wilson was a businessman from New Hampshire who sold the Army supplies during the War of 1812. The supplies were shipped in barrels that were marked "U.S." for the United States government. Just like today, people liked to invent nicknames, so the letters U and S evolved into the name "Uncle Sam." By the way, Uncle Sam's appearance is the product of artistic imagination; Mr. Wilson didn't look anything like our Uncle Sam.

 Did You Know
Building Codes

Most local governments have a set of unique building codes, but chances are good that these codes are based on the nation's Model Building Codes. Groups that know a lot about construction write these codes. In the past, there were three different national building codes. Today, the International Building Code has been adopted by all states except California, which has adopted a different one. Local governments usually start with these codes, then add their own sets of requirements. These codes are used to help protect the public. Building codes can dictate the strength of the structure, the placement of the structure on the lot, the type of electrical wire that can be used, and even how much distance there can be between porch uprights.

Sandy mentioned that building codes can affect housing costs, and she's right. In 1992, Hurricane Andrew swept through southern Florida, destroying virtually every building in its path. This brought about much stricter building codes for Florida, forcing building contractors to build much stronger houses. All of this costs money and may add thousands of dollars to the price of a new home.

Economics—Supply and Demand

Travis said that after the hurricanes passed over southern Florida in 2005, the cost of building supplies went up. He doesn't know how right he was! Think about it for a minute. Suppose you're at a major league baseball game, and you have the urge for a hot dog. Did you ever notice that a hot dog at a ballgame costs a lot more than one from a street vendor? Anyone who wants a hot dog while watching the ballgame needs to buy it from one of the vendors in the stadium. And the vendors know it, so they charge a lot for that hot dog! That's supply—a limited number of food vendors in the stadium—and demand—thousands of hungry people in the stadium.

Now think about a house with roof damage from a hurricane. The homeowner would probably pay just about anything to get plywood to fix it—that's demand. Smart retailers take advantage of this and try to send bigger shipments of building supplies to areas that have been damaged by storms.

Careers in Construction—Accounting

There are many different accounting jobs in the construction industry. Some require college degrees, while others may require no formal education at all. But since they deal with money, they all require you to be good at math—especially decimals!

Most companies use accounting programs to keep track of money, so everyone in accounting needs to know how to use these programs. Accounting clerks are entry-level accounting professionals (entry level usually means the position does not require a lot of experience), but they could make $10 to $22 an hour, depending on experience and location. Most companies want accounting clerks with some college education, but sometimes high school accounting is enough. Accounting clerks may keep track of material invoices, payroll, billing, or other items related to money.

Accountants have more responsibility than accounting clerks. Accountants usually need a four-year college degree. They might track overall costs for a project and help their employers predict a project's costs.

Certified Public Accountants (CPAs) are licensed by the state they work in and can fill out tax returns at the state and local level. Sometimes, CPAs need to sign a company's financial reports to certify that they are correct—even if an accountant or accounting clerk does them. CPAs have a lot of responsibility and can go to jail if they knowingly mislead the public with false financial statements.

NOTES

Chapter 4
Cattle Country

Surveying is an exacting skill that is best demonstrated using a hands-on approach. You may be able to borrow survey equipment to demonstrate these concepts, but you can also simulate survey equipment with a few easy-to-find components. For example, you can simulate a builder's level with a rifle scope or binoculars. If these are not available, try making one from an empty paper towel roll. A leveling rod can be simulated by hand-lettering measurements on paper and taping it to a two by four.

The 3–4–5 rule is more easily demonstrated than explained, so you may find that building a visual aid, such as the one shown here, is a good investment as a teaching tool. To demonstrate the 3–4–5 rule on a larger scale, go outside and set up strings to represent a building foundation and diagonal strings from the corners. When the corners are square, the diagonal strings will be of equal lengths. Set the foundation strings so the corners are not square—that way the diagonals will be unequal. (If you can't go outside, you can demonstrate this in the classroom using a corkboard and pushpins to hold the string.)

Have the students discuss what could happen if a survey was completed improperly. What would be the result if someone built a house too close to his or her neighbor's property line (setback violation)? Or worse, what if someone built on someone else's land?

NOTES

Chapter 4

Cattle Country

"This place is in the sticks," Phil exclaimed as the school van rolled down the sand road to the Browns' home site. Finally, someone wanted the class to build a house. The Browns came to Mr. Whyte just last week. The builder they had selected backed out of the job because the building site was so remote. And it was! It was in rural Alachua County, about a half hour drive from the school.

"Hey, it's not that bad," Al said, pointing to a narrow drive. "I live a mile down that road."

"It looks like Harry Potter's Forbidden Forest," Sandy said, looking at the Spanish moss hanging from the trees.

"Here's the Browns' place," Mr. Whyte said as he pulled up next to a white pickup truck with the words "Thomas Burke, Professional Land Surveyor" printed in red on the door. A man leaned against the truck with his arms crossed. He wore a baseball cap and sunglasses to protect him against the bright Florida sun.

"Don, nice to see you." The man called. Mr. Whyte slid out of the van and shook the man's hand. "So this is your crew. Hope you kids don't mind walking."

"Guys, this is Mr. Burke. He's a land surveyor. He's already finished a survey of this property but he's going to walk us through how it's done."

"I did this survey last week, and I used electronic equipment, but since you folks need to learn math, I'm going to show you how we did surveys before computers. There's a lot of math in a survey." Mr. Burke laughed a little when the class let out a groan. "You didn't think Mr. Whyte was going to let you get away without doing math, did you? We were college roommates—the man talks math in his sleep."

"I have always suspected this," Al said. Laughter erupted from the rest of the class.

"First things first." Mr. Burke continued. "Surveys are legal descriptions of property. They must be done with great accuracy—that's why a professional land surveyor needs to do them. If there's a dispute over boundaries, I need to prove my survey is correct before a judge, so I'm very careful."

Mr. Burke pulled a notebook from the back seat of his truck. "The first thing you need to understand is **azimuth**. Azimuth is direction; it's measured in degrees just like a circle. A circle is 360 degrees and when I divide the circle into quarters, each quarter has an angle of 90 degrees."

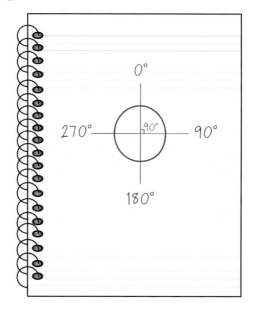

"Azimuth is also measured in 360 degrees. Zero degrees is always north, and north is shown with an arrow like this."

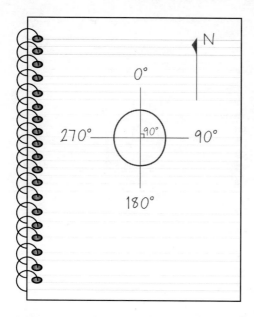

Mr. Burke turned the page and drew a simple compass. "When you face north, east is on your right, west is on your left, and south is behind you."

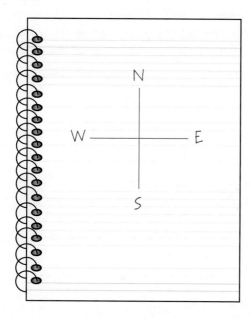

"Now I'll put the degrees back in, and you'll see that north is 0 degrees, east is 90 degrees, south is 180 degrees, and west is 270 degrees."

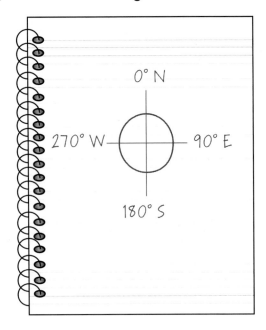

"So all we need to do is remember the direction and degrees and we've got it made," Travis said.

"For azimuth, yes," Mr. Burke said. "But surveyors use **bearing** for directional measurements. Bearing uses the angle from the north-south reference line; bearing can only have an angle of 0 to 90 degrees. For example, you face north and you turn to the east 45 degrees. The azimuth is 45 degrees and the bearing is north 45 degrees east, because you are 45 degrees east of north." He wrote in his notebook.

"Okay, Mr. Burke," Phil said. "I see what you're saying, but what about when you turn west? How do you measure the bearing then?"

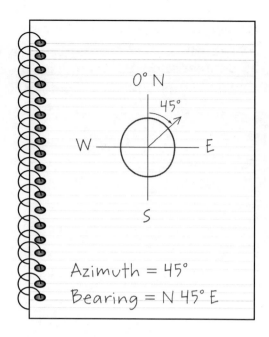

"If you're facing north and you turn to the west 45 degrees, the azimuth is 315 degrees. The bearing is north 45 degrees west, because you are 45 degrees west of north."

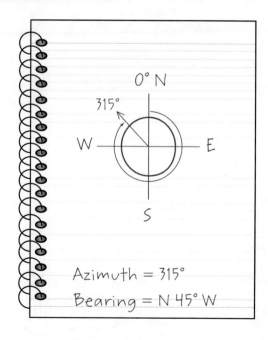

"Now look at what happens when you face south," Mr. Burke said. "When you turn to the east 45 degrees, your azimuth is 135 degrees and your bearing is 45 degrees east of south, so it's written south 45 degrees east."

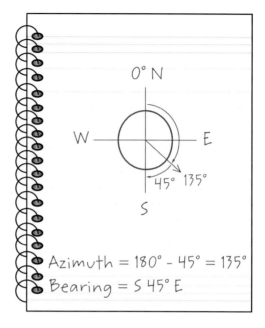

"Now face south and turn to the west 45 degrees. Your azimuth is 225 degrees and your bearing is 45 degrees west of south, so it's written south 45 degrees west."

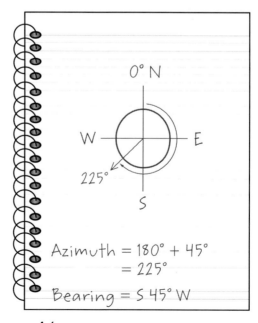

Math Speak

Azimuth is a directional measure of 0 to 360 degrees. North is both 0 degrees and 360 degrees. East is 90 degrees, south is 180 degrees, and west is 270 degrees.

Bearing is an angular direction too, but it uses the north-south line as a reference and can only be between 0 and 90 degrees.

Bearing measurements are given in three pieces:

1. Reference line (north or south)
2. Degrees from reference
3. Direction from reference (east or west)

A bearing of N 42 degrees W is interpreted as from a north direction, turn 42 degrees to the west.

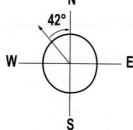

A bearing of S 42 degrees W is interpreted as from a south direction, turn 42 degrees to the west.

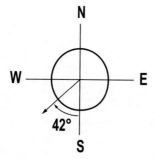

Practice Problems 4-1

For Questions 1 through 5, convert the azimuth given to its corresponding bearing.

1. 95 degrees = **S 85 degrees E**
2. 185 degrees = **S 05 degrees W**
3. 200 degrees = **S 20 degrees W**
4. 275 degrees = **N 85 degrees W**
5. 349 degrees = **N 11 degrees W**

For Questions 6 through 10, draw a figure like the one shown here and then show the direction indicated by the given bearing:

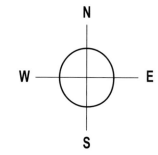

6. N 58 degrees E

   N 58° E

   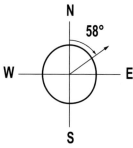

7. N 27 degrees W

   N 27° W

   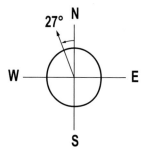

8. N 10 degrees E

   N 10° E

   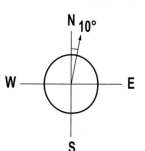

9. S 25 degrees E

   S 25° E

   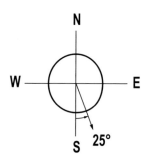

10. S 47 degrees W

    S 47° W

    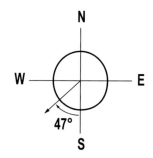

"Here's a copy of the survey I did last week. We'll walk the boundaries and you can ask me some questions."

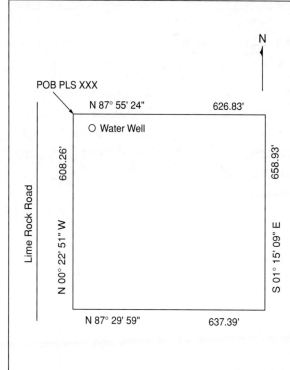

LEGAL DESCRIPTION (9.185 ACRES):
A PARCEL OF LAND SITUATED IN THE SOUTHEAST QUARTER (1/4) OF THE NORTHEAST QUARTER (1/4) OF SECTION 29, TOWNSHIP 7 SOUTH, RANGE 20 EAST, ALACHUA COUNTY, FLORIDA BEING MORE PARTICULARLY DESCRIBED AS FOLLOWS:

FOR A POINT OF REFERENCE, COMMENCE AT THE NORTHWEST CORNER OF THE SOUTHEAST QUARTER (1/4) OF THE NORTHEAST QUARTER (1/4) OF SECTION 29 AND RUN SOUTH 00 DEG. 22 MIN. 51 SEC. EAST, A DISTANCE OF 494.84 FEET TO THE POINT OF BEGINNING OF THE HEREINAFTER DESCRIBED PARCEL; THENCE RUN NORTH 87 DEG. 55 MIN. 24 SEC. EAST, A DISTANCE OF 450.83 FEET TO A POINT, SAID POINT BEING 0.30 FEET SOUTH OF A FENCE CORNER; THENCE CONTINUE NORTH 87 DEG. 55 MIN. 24 SEC. EAST PARALLEL WITH AND 0.30 FEET SOUTH OF A WIRE FENCE, A DISTANCE OF 176.00 FEET; THENCE RUN SOUTH 01 DEG. 15 MIN. 09 SEC. EAST, A DISTANCE OF 658.93 FEET; THENCE RUN NORTH 87 DEG. 29 MIN. 59 SEC. WEST, A DISTANCE OF 637.39 FEET; THENCE RUN NORTH 00 DEG. 22 MIN. 51 SEC. WEST, A DISTANCE OF 608.26 FEET TO THE POINT OF BEGINNING.

CONTAINING 9.185 ACRES, MORE OR LESS

"Um...Mr. Burke," Jorge began, looking at the paper Mr. Burke handed out. "The numbers are written in degrees, feet, and inches. I never heard of that."

"I forgot to explain that. Bearings on a survey aren't measured in decimals or fractions. They're measured in minutes and seconds. The abbreviations for minutes and seconds look just like feet and inches." Mr. Burke pointed to the top boundary line on the survey. "It says N 87 degrees, 55 minutes, 24 seconds."

"What?" Al said. "We just learned decimals and fractions. Now there's minutes and seconds? This is too much. I demand to speak to the man in charge."

"Al's our resident comedian," Mr Whyte told Mr. Burke. "Keep it down, Al. You're embarrassing me."

Mr. Burke laughed. "Sorry to break the news to you, Al, but it's true. Minutes and seconds are easy to remember because there are 60 minutes in a degree and 60 seconds in a minute—just like telling time."

Mr. Burke brought out his notebook again. "All you need to do is take the decimal part of the degree measurement and multiply it by 60 to get minutes. So if I measure 10.5 degrees, I take the 0.5 and multiply it by 60 to get 30 minutes. So 10.5 degrees equals 10 degrees and 30 minutes."

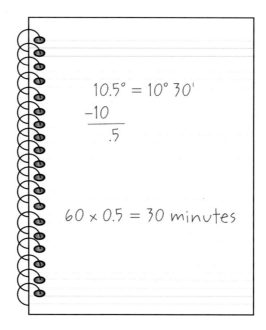

"If I measure 10.55 degrees, I take the 0.55 and multiply it by 60 to get 33 minutes, so 10.55 degrees converts to 10 degrees and 33 minutes."

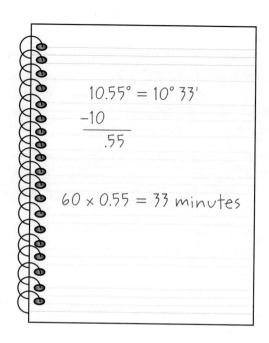

"If I measure 10.555 degrees, I take the 0.555 and multiply it by 60 and I get 33.3 minutes. When you have a decimal in the minutes, you need to convert it to seconds, so I take the 0.3 and multiply it by 60 to get 18 seconds. So 10.555 degrees converts to 10 degrees, 33 minutes, 18 seconds."

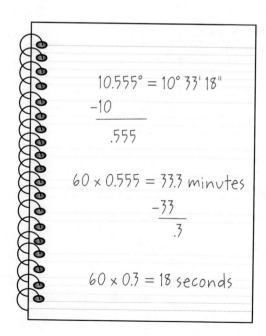

Key To Understanding

Bearing is given in degrees/minutes/seconds format. There are 60 minutes in one degree and 60 seconds in one minute, so there are 3,600 seconds in one degree (60 minutes/per degree times 60 seconds/per minute = 3,600 seconds per degree).

To convert angular values given in decimal degrees into equivalent values of degrees, minutes, and seconds, use the following procedure. As an example, we will convert 20.9964° to degrees, minutes, and seconds, rounded to the nearest second.

Step 1 Subtract the whole number of degrees.

$$20.9964° - 20° = 0.9964°$$

Step 2 Multiply the remaining decimal value by 60 to convert it into minutes.

$$60 \times 0.9964 = 59.784'$$

Step 3 Subtract the whole number of minutes.

$$59.784' - 59' = 0.784'$$

Step 4 Multiply the remaining decimal value by 60 to convert it into seconds.

$$60 \times 0.784 = 47.04"$$

Step 5 Combine the results.

$$20.9964° = 20° \, 59' \, 47"$$

Angular measurements expressed in degrees, minutes, and seconds can be converted to decimal degrees by using the following formula:

Decimal degrees = degrees + (minutes ÷ 60) + (seconds ÷ 3,600)

For example, to convert 20° 59' 47" to decimal degrees, proceed as follows:

Decimal degrees = 20 + (59 ÷ 60) + (47 ÷ 3,600)

Decimal degrees = 20 + 0.983333333 + 0.013055556

Decimal degrees = 20.996388889 = 20.9964°
(rounded off to 4 decimal places)

"Let's walk the boundaries," Mr. Burke said. "We're going to convert degrees, minutes, and seconds to decimal as we go, so everyone bring your notebook and calculator."

"We can't use calculators," Tom said with a grin. "We live in a fantasy world."

"Only in the classroom, Tom," Mr. Whyte said with a smile. "This is the real world. I put a box of calculators in the van. Everyone get your hats and sunglasses. Sleeves stay down and buttoned—I don't want to see any sunburns this afternoon. Carry your gloves with you—you never know when you'll need them. Travis, do you have your bee sting kit?"

"Yes, sir." Travis waved a small red plastic case.

"Who's your buddy?"

"Olivia. She's the only one who's not afraid of needles."

"Good. Olivia, I'm counting on you." Olivia nodded. Her little brother was allergic to bee stings, so she knew what to do if Travis got stung.

"Let's go." Mr. Burke took off walking. "We're moving north to the northwest corner of the property. Everyone hold your survey so that the north arrow is facing north. See how Lime Rock Road is shown on the left side of the survey and is on our left hand side? Here's the well—it's shown on the survey too. And this is the point of beginning." Mr. Burke squatted near a fence post and parted some grass to reveal a survey marker. "I put this marker here. It marks the corner of the property. When you first go to a job site, make sure you know where these markers are. Legally, you should never cross a boundary to someone else's property without permission—it's trespassing." The class quickly backed away and he smiled.

"Look at the survey. In the upper left corner, you see this marked as the point of beginning or POB. The top boundary says 87 degrees, 55 minutes, and 24 seconds. Let's step into the shade and everyone convert that number to decimal degrees."

"Why do we have to convert it? Why can't we just use it as it is?" Tom asked.

Mr. Whyte smiled. "Because this is a class, Tom, and you're learning how to use degrees, minutes, and seconds. You learn by doing, so do it!"

For a few minutes, the students studied the numbers. Phil and Tom whispered to each other. Jorge leaned towards them to add to the conversation. They punched buttons on calculators, wrote in their notebooks, and then compared notes. Here's what they calculated:

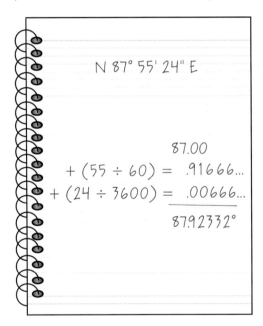

"It's unanimous, Mr. Burke, we all came up with 87.92332 degrees." Al said.

"Good. I'm facing north and I'm going to use my compass and turn to the east almost 88 degrees, and walk to the next corner. How many feet is it to the next corner?"

The students looked at their copies of the survey. "Looks like 626.83 feet."

"Let's go." The class trailed after Mr. Burke. Mr. Whyte dropped to the back of the crowd and fell into step with Sandy.

"I know you have a lot on your mind, Sandy, but you need to concentrate."

"I know, Mr. Whyte, but it's really hard. I keep wondering what's going on at home."

Mr. Whyte nodded. "I know, but if you can focus on class, it'll push the home situation out of your mind and you'll feel better."

Sandy gave Mr. Whyte a weak smile. "I'll try."

"Go on up and talk to Al and Tom. If anyone can take your mind off your troubles, it's those two." Sandy gave a little laugh and ran to catch up with Al and Tom.

Practice Problems 4-2

Mr. Burke had the students convert the degrees, minutes, and seconds into decimal degrees as they walked the property. Convert the following measurements to decimal degrees and then to azimuth, and then trace the property boundaries on the survey.

1. South 01° 15' 09" East = **1 + (15 ÷ 60) + (9 ÷ 3600) =**

    South 1.2525 degrees East = 178.7475 degrees azimuth

2. North 87° 29' 59" West = **87 + (29 ÷ 60) + (59 ÷ 3600) =**

    North 87.4997 degrees West = 272.5003 degrees azimuth

3. North 00° 22' 51" West = **0 + (22 ÷ 60) + (51 ÷ 3600) =**

    North 0.3808 degrees West = 359.6192 degrees azimuth

Instructor Note: Answers may vary slightly due to differences in rounding.

After the students walked the boundaries of the Browns' property, they helped Mr. Burke unload some equipment from his truck—a builder's level, a tripod, and a leveling rod. Mr. Burke took a stack of books out of the back seat of his truck, and then placed a tape measure, plumb bob, and string level on top of the books.

"Look," Mr. Burke said. "Here's a picture of a survey crew using old equipment like this."

"Are you in this picture, Mr. Whyte?" Al asked.

"Funny, funny, Al. Let's get out of the sun," Mr. Whyte said. "Tom, Phil, be very careful with those instruments. They're delicate."

Mr. Burke waited until the group moved the equipment and then settled down under a stand of pines. It was cooler out of the sun and a gentle breeze blew through the trees.

"Surveying requires accuracy and precision," Mr. Burke began. "Accuracy is how close the measurements I make are to the real measurements. Precision is my degree of error when I measure the same distance over and over again. For example, if I measure from point A to point B three times and I get the exact same measurement each time, I'm precise, but not necessarily accurate—I could be measuring it wrong each time. If I have three of you guys measure it and we all get the same measurements, I'm probably accurate as well as precise."

Mr. Whyte pulled out his survey again. "Remember how I said that a property survey is a legal document? Well, the law takes a dim view of inaccurate, imprecise surveys. So everything I do needs to be exact. I need to start my measurements in exactly the right spot, and I need to make accurate and precise measurements.

Before computers were around, it could take days to survey this property, and I would have had at least one other person helping me, probably two or three. With electronic equipment, I was able to do this survey by myself in a few hours."

"The property owners marked the general location of the house," Mr. Whyte said. "And Mr. Burke marked the exact location of the corners. Now he's going to show us how to measure the distance from the property boundaries to the house. This is called **setback**. Each local government sets the minimum number of feet that a building must be located from the property line. The Browns want their house in the middle of the property, so setback in this case is a couple hundred feet. That's well within the county limit. If we were building a house in town, space is limited and the code is much tighter."

"One important thing to remember about surveys is that you measure the distances as if there are no hills or valleys," Mr. Burke said. "When you have a long distance and a big hill, this can be a big problem." Mr. Burke drew on his notebook and held it up for the class to see.

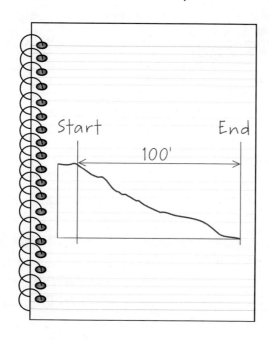

Key To Understanding

Accuracy is when a measurement is very, very close to the actual measurement. Precision is being able to get the same answer each time you make the same measurement.

(A) PRECISE BUT NOT ACCURATE

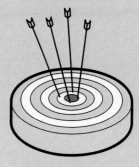

(B) ACCURATE BUT NOT PRECISE

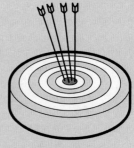

(C) BOTH ACCURATE AND PRECISE

"How are you supposed to measure something like this? The distance is long, and the hill is steep. There's no precision or accuracy in trying to measure this distance at one time, so what we need to do is called breaking the tape. I brought some books about surveying, so I have a picture of it." Mr. Burke opened a book and showed the class a picture.

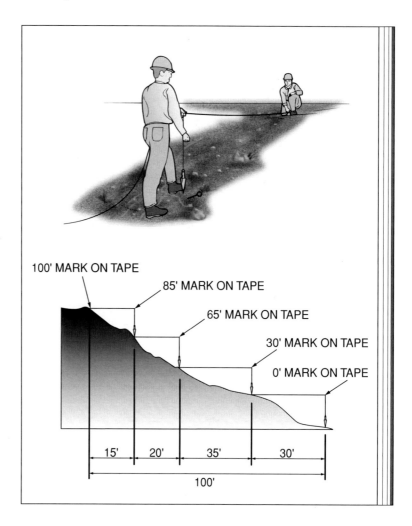

"Two people hold the tape and carefully measure a short distance. There's a plumb bob attached to the end of the tape to precisely mark the start of the next measurement. The people move down the hill, measuring short pieces as they go. After the measurement is complete, all of the pieces are added to get the total distance."

"No wonder it took so many people to do a survey," Sandy said, looking around the property and scribbling in her notebook. "Mr. Burke, this place is really big. Each boundary is over 600 feet, so the area is more than 360,000 square feet."

"That's right, Sandy." Mr. Whyte said. "When you have this much property, it's measured in acres. One acre equals 43,560 square feet. This property is a little over 9 acres."

"How do you figure out the area?" Sandy asked. "The property is irregularly shaped, the corners aren't 90-degree angles. It's not a **parallelogram**. It's not a trapezoid. It looks impossible."

"It's not. You actually have everything you need on the survey. You just need to know a little…"

"Wait," Al interrupted, holding his hand up in a stop signal. "Let me guess. We need to know a little more math."

Mr. Burke laughed. "You got it Al, but don't worry—I'm not going to make you find the area of this property. Come on, let's find the markers for the house corners."

Mr. Burke showed the class how to attach the plumb bob to the tape measure and use a string level to measure from the west boundary to where the Browns wanted their house. Travis and Al held the measuring tape tight while Sandy placed the string level on the tape. She looked at the bubble and told Travis to move his end of the tape up until it was level. Jorge marked the place on the ground indicated by the plumb bob. Olivia wrote the first measurement in her notebook.

 Mind Games

Area is the amount of space on a surface. It is measured in square units: square inches, square feet, and square yards. Large pieces of land are measured in acres and sometimes even square miles. Since one acre always equals 43,560 square feet, you don't say square acres—the measurement is just called an acre.

The formula to find area is always the same:

Length × Width = Area

When you need to find the area of a square, you can use the formula s^2 (s is for side) because the length and width of a square are the same.

Al moved to the place Jorge marked on the ground and placed his end of the tape in the precise location. Travis stretched the tape between them again. The class worked their way through the trees to the open field until they finally came to a marker in the field. Mr. Burke had placed the marker when he did the survey. Olivia added up all the measurements. It was 225 feet east of the west boundary.

Mr. Whyte carried some wooden stakes and a rubber mallet to the home site. "I'm going to use these to mark the corners of the house. I need each corner to be a 90-degree angle so the foundation will be true." Mr. Whyte held up three wooden stakes that had been nailed together with two wooden slats to form a corner. "This is a batter board, and it is used to mark the corners of a building."

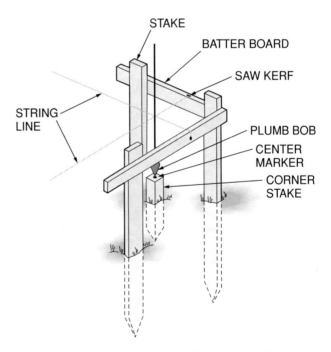

Mr. Whyte lightly tapped the three wooden stakes into the ground behind the surveyor's stake. "I'm placing these stakes shallowly until I'm sure I have a 90-degree angle."

"Do we need a square, Mr. Whyte?" Phil asked, shading his eyes from the glare of the sun. "I can run back to the van and get one."

"No, Phil, but thanks for asking. I'm going to show you a trick of the trade. It's called the 3–4–5 rule of right triangles. It doesn't need any fancy tools or instruments. All you need are some wooden stakes and slats, string, and a rule. First, the 90-degree corner." He pulled out his notebook.

"The 3–4–5 rule of right triangles says that if you make a 90-degree angle with one side being 3 feet and the other being 4 feet, when it's closed in to make a triangle, the third side must be 5 feet. This rule has been used for centuries."

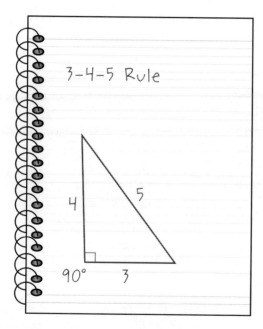

"How do you know that's right, Mr. Whyte? People used to believe the world was flat and now we know it's round," Jorge asked.

"Glad you asked, Jorge. Let's move into the shade and talk about it." When everyone was settled under the pine trees, Mr. Whyte continued. "Take your protractor

Math Speak

A right triangle is a triangle with one 90-degree angle. The 3–4–5 rule of right triangles says that if the sides that make the right angle are three units and four units, then the other side must be five units.

3–4–5 Rule

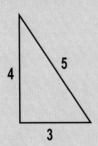

and mark a 90-degree angle in your notebook. Then use your ruler to make a right angle. Start at the bottom left corner of the page, because I want you to draw a couple of triangles. Draw a 90-degree angle with one line 3 inches long and another 4 inches long. Now connect the two and measure the length of the third line. If you measured accurately, the third line is 5 inches. That's the 3–4–5 rule in action."

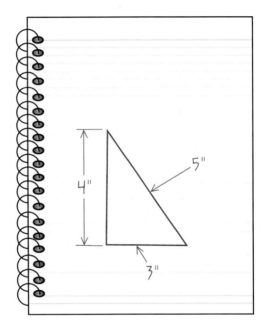

"Now let's do a multiple of two—3 inches times two is 6 inches and 4 inches times two is 8 inches. So the 3–4–5 rule of right triangles says the third side must be 5 inches times two or 10 inches."

"Hey, cool," Travis said. "It is 10."

Mr. Whyte continued. "The 3–4–5 rule is only good for some factor of these dimensions, and all triangles don't necessarily have these dimensions. We could want a triangle with different measurements." He turned to a new page in his notebook.

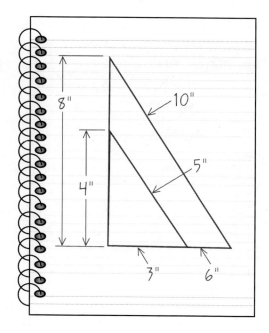

"Some enterprising soul discovered long ago that he could calculate the lengths of the sides of right triangles with the formula a² plus b² equals c². The trick to remember is that the long side is always c." Mr. Whyte wrote in his notebook and showed the class.

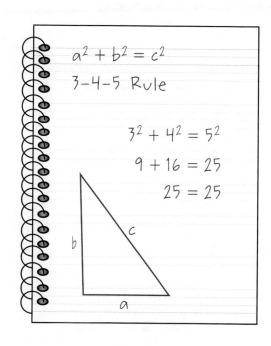

"Now, let's prove it. We'll use the millimeter side of your rulers this time. Draw a right angle with one side 50 millimeters long and the other 70 millimeters, and using the a^2 plus b^2 equals c^2 formula, let's solve for c. And yes, Tom, you may use a calculator!"

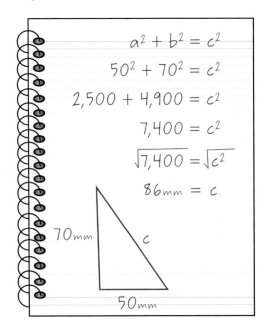

"After you solve for c, I want you to measure the length of c with your ruler."

"I got 86 millimeters when I solved for c and it measures pretty close to 86 millimeters with a ruler," Olivia said. The others nodded their heads.

"Good. By the way, this is called the Pythagorean theorem."

Key To Understanding

The 3–4–5 rule of right triangles is based on the Pythagorean theorem, which was developed by an ancient Greek mathematician named Pythagoras. The Pythagorean theorem is:

$$a^2 + b^2 = c^2$$

Side c is always the long side of the triangle.

Using the Pythagorean theorem, you can calculate the length of the third side if you know the lengths of the other two sides.

 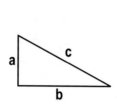 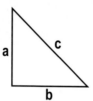

Know a and b, calculate c:

$a^2 + b^2 = c^2$

$\sqrt{a^2 + b^2} = \sqrt{c^2}$

$\sqrt{a^2 + b^2} = c$

Isolate c by taking the square root of both sides of the equation. To find the square root of a number, you must find the number that when multiplied by itself equals the first number. The square root of 16 is 4 because 4^2 equals 4×4 or 16.

Know b and c, calculate a:

$a^2 + b^2 = c^2$

$a^2 + b^2 - b^2 = c^2 - b^2$ Isolate a by subtracting b^2 from both sides of the equation.

$a^2 = c^2 - b^2$

$\sqrt{a^2} = \sqrt{c^2 - b^2}$ ← Then take the square root of both sides of the equation.

$a = \sqrt{c^2 - b^2}$

Know a and c, calculate b:

$a^2 + b^2 = c^2$

$a^2 + b^2 - a^2 = c^2 - a^2$ Isolate b by subtracting a^2 from both sides of the equation.

$b^2 = c^2 - a^2$

$\sqrt{b^2} = \sqrt{c^2 - a^2}$ ← Then take the square root of both sides of the equation.

$b = \sqrt{c^2 - a^2}$

Practice Problems 4-3

Using the Pythagorean theorem, calculate to the second decimal place the length of the unknown side of the following right triangles.

1.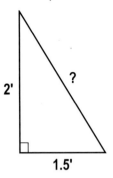

$a^2 + b^2 = c^2$
$2^2 + 1.5^2 = c^2$
$4 + 2.25 = c^2$
$6.25 = c^2$
$\sqrt{6.25} = \sqrt{c^2}$
$2.5' = c$

2.

$a^2 + b^2 = c^2$
$8^2 + 5^2 = c^2$
$64 + 25 = c^2$
$89 = c^2$
$\sqrt{89} = \sqrt{c^2}$
$9.43' = c$

3.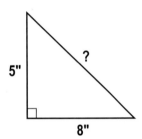

$a^2 + b^2 = c^2$
$a^2 + 5^2 = 7^2$
$a^2 + 25 = 49$
$a^2 + 25 - 25 = 49 - 25$
$a^2 = 24$
$\sqrt{a^2} = \sqrt{24}$
$a = 4.9'$

4.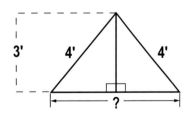

$a^2 + b^2 = c^2$
$3^2 + b^2 = 4^2$
$9 + b^2 = 16$
$9 - 9 + b^2 = 16 - 9$
$b^2 = 7$
$\sqrt{b^2} = \sqrt{7}$
$b = 2.65'$
$? = 2.65' + 2.65'$
$? = 5.3'$

"Now back to setting the corners of the house," Mr. Whyte said. "You probably all agree that we want the corners of the house to be 90 degrees. And using the Pythagorean theorem, we can figure out the length of the third side when we know the length of the other two sides." He took one of the batter boards and put it down by a stake.

"Mr. Burke placed stakes that mark the exact corners of the house. I'm placing these batter boards outside the area of the house so we have room to work inside with tools and heavy equipment." Mr. Whyte set his batter boards and used the 3–4–5 rule to square the sides to a 90-degree angle, then he pounded the stakes deep into the ground.

"Okay, guys. Teams of three people," Mr. Whyte continued. "Everyone find a corner stake and set the batter boards about three feet from the corner stake."

The students did as they were told and set the batter boards and then squared them so they were all at 90-degree angles.

"Mr. Whyte, this doesn't seem accurate to me," Tom said. "We set the batter boards outside the area of the house, and sure, they're set to 90-degree angles. But how can we use them when they're so far from the foundation?"

"Good question, Tom. We're going to use string to mark the real corners." Mr. Whyte picked up a ball of white nylon string and tied one end of the string around the batter board closest to him. He handed the ball of string to Tom. "Go down to the next batter board and tie the string around it, just like I did. Tie it tight so it doesn't sag."

Mr. Whyte helped the students tie the string around each batter board to form a corner. Mr. Whyte and Mr. Burke moved the strings on the batter boards until the first string corner was directly over the house's corner stake.

Then Mr. Whyte attached a plumb bob to the string corner and adjusted the strings until the plumb bob was exactly over the center mark in the corner stake. He used a staple gun to attach the strings so they couldn't slide around on the batter board. Mr. Burke did the same to the strings on the opposite corner.

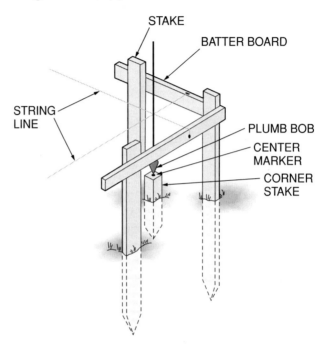

"These strings represent the footprint of the foundation," Mr. Whyte said. "You guys go over and make sure the strings meet over the corner stakes. Use the plumb bob to be sure it's exact."

When they finished adjusting the strings to mark the foundation, Mr. Whyte showed the class how to run diagonal strings from corner to corner. They knew that each corner was set to 90 degrees, because the diagonals were both the same length.

"Good job, people. Let's break for lunch."

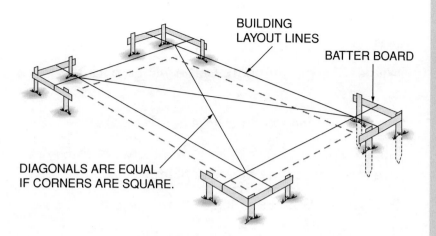

After lunch, the students clustered around Mr. Burke and Mr. Whyte. Mr. Burke talked about the house site. "Did anyone notice the lay of the land at the house site? It slopes downhill from west to east. How are you guys going to make sure the slab is level?"

"Um, Mr. Whyte," Al said, looking over his shoulder into the field. "Do the Browns have cows?"

"I don't think so. Why do you ask?"

Al pointed to the field. "I think we might have a problem."

Mr. Whyte looked in the direction Al was pointing. "What…"

"I'll be," Mr. Burke said. "Cattle."

The group turned and saw a dozen or so animals roaming the field, eating grass. One was a pale tan color and had wide horns.

"That is one ugly cow," Sandy said.

"That's not a cow, Sandy. It's a steer," Al said. "It's raised for beef."

"Hey, Al, that one will look good in your convertible," Tom laughed.

"Um, Al," Mr. Whyte said. "I'm a city boy. You got any ideas on what we should do?"

"We need to run them off. Cattle will go back the way they came."

"Are you sure they won't charge us?"

"No, sir. I'm not, but chances are pretty good they're more afraid of us than we are of them."

Mr. Whyte looked at Mr. Burke. "I hate when people say that."

"Come on, Al, looks like it's up to us," Mr. Burke said as he jogged toward the animals. "Surveyors are used to cattle."

Mr. Burke and Al trotted toward the cattle, shouting and waving their arms. "Git on you. Haw! Haw!" The cattle took off toward the north boundary of the property. Once they all got going, a thunder sound filled the air. Soon everyone joined in the chase. "Haw. Haw!" One small brown steer dodged Travis, and chased Mr. Whyte around the field until Sandy, Phil, and Tom surrounded it and turned it toward the rest of the herd, running north.

"I didn't know you were cowboys." Mr. Whyte laughed.

"Neighbor's fence must be down," Al said. "I help my uncle with his cattle sometimes, and one thing's for sure—they'll be back. Cattle are like that, they follow the same path all the time."

"I'll have to call the owners and see what they want me to do," Mr. Whyte said. "Okay, back to work. Mr. Burke was talking about the slope of the house site. We want a nice level foundation, so we need to prepare the site."

"I'm going to show you how to measure the elevation of a site," Mr. Burke said, picking up the tripod. "Al, you and Sandy bring that box and follow me. Phil, grab the leveling rod—you can be the rod man."

"I have to set up the tripod exactly level, because I'm going to mount the builder's level on it. A builder's level is like a telescope. I set it on the tripod and look through it at the rod man. The rod man holds the leveling rod straight and level and I look through the eyepiece of the builder's level to get an elevation reading. I'm cheating a little because I've already done the survey and know the elevation at this corner of the house is 147 feet. So I'm using it as a reference." Mr. Burke explained that to a surveyor, elevation is the distance above sea level. Then he carefully set the tripod between two corner stakes. Mr. Burke adjusted the tripod until it was stable. Then he adjusted the leveling screws on the tripod until he was sure it was level.

Mr. Burke told Phil to hold the level next to the corner stake. "I already know the elevation at that corner is 147 feet, so I'm going to use that as my reference." Mr. Burke looked through the eyepiece. "I read 6.25 feet on the rod. Since I know the elevation is 147 feet, then my instrument height must be 147 plus 6.25 or 153.25 feet. That's called the backsight reading. Phil, you need to go down to the east corner. Be really careful when you walk by me, because if you bump the instrument, we need to start all over again."

Phil went to the other corner stake and held the rod straight and level. Mr. Burke loosened a screw and slowly turned the builder's level to face Phil. When Mr. Burke was satisfied that the builder's level was set properly, he looked through it and said, "I have a reading of 7.25 feet. That tells me that this corner is one foot lower in elevation than the first corner."

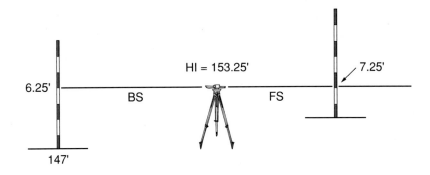

Mr. Burke let everyone look through the eyepiece to sight the leveling rod and together they measured the elevation of each corner of the house. He showed them how to set stakes to show different elevation levels. Mr. Whyte said that they would use those stakes to make sure the house slab was level. He had everyone write the numbers in their notebooks, because they would need the measurements to calculate excavations for the foundation.

A red Ram pickup turned into the property while the class helped Mr. Burke load his equipment into his truck. Tom stood still and stared at the truck.

"Oh, good," Mr. Whyte said, turning to the class. "Since this job site is so far from the school, I'm not going to be here as much as I'd like, so JM Custom Homes is donating the services of a construction foreman. This must be him."

"Oh, no," Tom said quietly. "It's my dad."

Chapter 4 Review Problems

1. Convert the following azimuth readings into bearing measurements:

    a. 17 degrees = **North 17 degrees East**

    b. 99 degrees = **South 81 degrees East**

    c. 180 degrees = **South 0 degrees**

    d. 330 degrees = **North 30 degrees West**

2. Convert the following bearing measurements into azimuth readings:

    a. South 25 degrees West = **180 + 25 = 205 degrees**

    b. North 90 degrees East = **0 + 90 = 90 degrees**

    c. South 80 degrees East = **180 − 80 = 100 degrees**

    d. North 30 degrees West = **360 − 30 = 330 degrees**

3. Convert the following measurements to decimal degrees:

    a. 05° 22' 30" = **5 + (22 ÷ 60) + (30 ÷ 3600) = 5.375 degrees**

    b. 08° 47' 03" = **8 + (47 ÷ 60) + (3 ÷ 3600) = 8.7842 degrees**

    c. 37° 15' 23" = **37 + (15 ÷ 60) + (23 ÷ 3600) = 37.2564 degrees**

    d. 81° 30' 58" = **81 + (30 ÷ 60) + (58 ÷ 3600) = 81.5161 degrees**

4. Convert the following measurements to degrees, minutes, seconds:

    a. 15.56 degrees = **15.56 − 15 = 0.56 × 60 = 33.6 − 33 = 0.6 × 60 = 36 = 15° 33' 36"**

    b. 32.256 degrees = **32.256 − 32 = 0.256 × 60 = 15.36 − 15 = 0.36 × 60 = 21.6 or 22 (rounded up) = 32° 15' 22"**

    c. 48.875 degrees = **48.875 − 48 = 0.875 × 60 = 52.5 − 52 = 0.5 × 60 = 30 = 48° 52' 30"**

    d. 62.455 degrees = **62.455 − 62 = 0.455 × 60 = 27.3 − 27 = 0.3 × 60 = 18 = 62° 27' 18"**

5. Use the Pythagorean theorem to calculate the distance between Point 1 and Point 2.

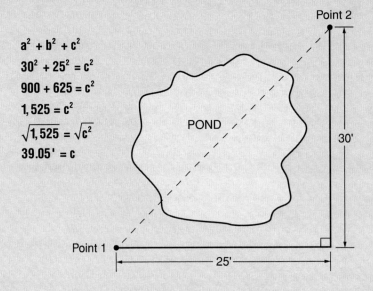

$a^2 + b^2 + c^2$
$30^2 + 25^2 = c^2$
$900 + 625 = c^2$
$1,525 = c^2$
$\sqrt{1,525} = \sqrt{c^2}$
$39.05' = c$

6. Use the Pythagorean theorem to calculate the distance between Point 1 and Point 2.

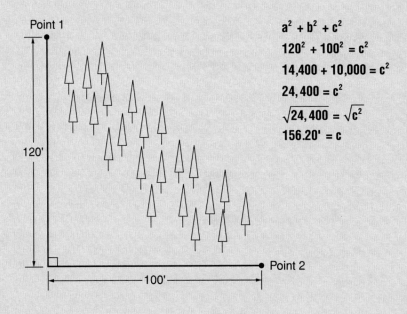

$a^2 + b^2 + c^2$
$120^2 + 100^2 = c^2$
$14,400 + 10,000 = c^2$
$24,400 = c^2$
$\sqrt{24,400} = \sqrt{c^2}$
$156.20' = c$

7. You are working on a construction job and need a ladder to reach the roof. The eaves of the roof are 32 feet off the ground. You want the ladder to extend 3 feet beyond the resting point for safety. You don't want the ladder to move while you're on it, so you are going to rest the bottom of the ladder against a brick wall that is 10 feet from the house. Calculate the required length of the ladder. (Hint: Draw a picture.)

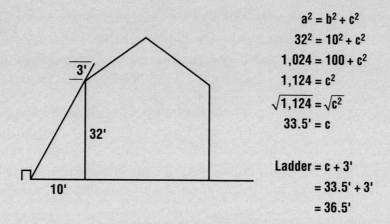

$$a^2 = b^2 + c^2$$
$$32^2 = 10^2 + c^2$$
$$1{,}024 = 100 + c^2$$
$$1{,}124 = c^2$$
$$\sqrt{1{,}124} = \sqrt{c^2}$$
$$33.5' = c$$

Ladder = c + 3'
= 33.5' + 3'
= 36.5'

8. Write three or more sentences stating the importance of a survey to a construction crew. (Hint: You could start with the words: "Surveys are important to construction crews because..." then you could write about what you think would happen if a survey was done incorrectly, or if the construction crew ignored the survey.)

Answers will vary, but may include:

Surveys are important to construction crews because they are legal documents that show ownership of land. Only owners can authorize construction on the property. If a survey is incorrect, it may cause the construction crew to trespass on another person's property, to build in the wrong area, or to inadvertently violate local ordinances such as setback. These errors can be costly, since the crew may need to demolish some or all of their work. Construction workers should never knowingly ignore survey markers. This can cause the same problems as those stated above.

 Terms

Azimuth: The horizontal direction of a position expressed in degrees from 0 to 360, using north as the fixed reference. Azimuth moves clockwise from north through 360 degrees.

Bearing: An angular direction using north and south as its reference. It can only be between 0 and 90 degrees.

Parallelogram: A four-sided figure with the opposite sides being parallel.

Setback: The minimum distance from a boundary line a structure must be located. Setbacks vary from municipality to municipality and maybe even from neighborhood to neighborhood. The reason for setback is to permit enough room to allow access to the back of the home, and to allow room to prevent the spread of fire to neighboring buildings.

 History—The Homestead Act

In 1862, Abraham Lincoln signed into law the Homestead Act. This act turned over a large amount of public domain land to the public. Anybody who was at least 21 years old could claim a 160-acre quarter-section of land. In exchange, the homesteader had to live on the land, build a home, and farm the land for five years. After the five years, the homesteader could "prove up" the claim and receive a patent (or deed) for the land. Then it was theirs to keep or sell as they pleased.

The cost for this? A $10 filing fee, $2 in commission to a land agent, and after five years, a $6 fee for the patent. It was hard work living off the land in wild, unsettled country and not everyone stuck it out, but for those who did, it meant owning 160 acres of land.

The Homestead Act was repealed in 1976, except for Alaska, where it stayed open until 1986. In total, 270 million acres, or 10% of the area of the United States, was claimed and settled under the Homestead Act of 1862.

Before homesteaders could claim the land, surveyors had to survey the land. They used the Public Land Survey System (PLSS) to describe parcels of land. In the PLSS, townships are 36 square miles in area and are divided into sections equal to one square mile. Each section contains 640 acres. Homesteaders were allowed to claim a quarter section or 160 acres. A typical township is shown on the next page.

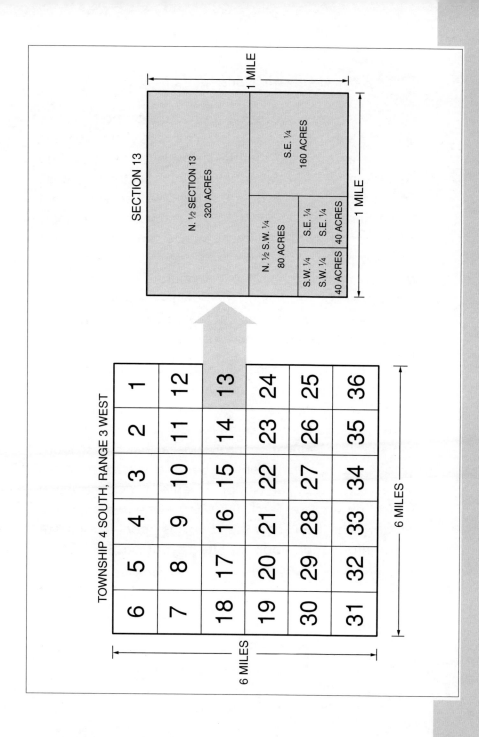

Careers in Construction—Survey Team

Only a licensed surveyor can perform a property survey and prepare a legal description of property, but there are many other job opportunities in surveying. People who help the surveyor may not have any formal training at all. They simply learn by doing. There are also site layout technicians. They know how to care for and use survey equipment, record field notes for the surveyor, and read and interpret construction drawings and specifications. More advanced technicians may operate equipment and make field note entries on their own—they may perform all the duties of a surveyor, but remember, only a licensed surveyor can prepare a legal document.

Members of survey teams need to be well organized and precise in their work, and—you guessed it—good at math! Some states don't require that licensed surveyors have a college degree, but most do. Survey technicians can enroll in education programs at various schools, academies, or training centers. Some of these institutions are accredited by the National Center for Construction Education and Research (NCCER)—when you see the NCCER logo, you know the course is good! If you want more information, look up Site Layout on the NCCER web site (www.nccer.org).

 **Did You Know**

Spanish moss is a common sight in warm, humid climates. This plant is not really moss at all. It's a plant that hangs from trees. When a lot of Spanish moss is hanging from a tree, it gives the tree an eerie look, but Spanish moss doesn't damage the tree. Spanish moss gets water and nutrients from rain and the air, so it is often called an air plant. Many orchids are air plants, too.

There are other plants that hang from trees, but these plants get their nutrients from the trees—these plants are called parasites, because they can damage or even kill the tree. Mistletoe is a parasitic plant.

NOTES

Chapter 5
Breaking Ground

Chapter 5 discusses volume. You can illustrate volume with any vessel. Fill a cup with water, sand, or flour. Ask your students to compare the weight of various substances—same volume, different weight. Take a pail of water and an empty soda bottle and push the bottle into the water. Even though the bottle is empty, the water will rise. Fill the soda bottle with sand and let it sink in the water. Explain that weight has nothing to do with volume.

Show the students that volume can take any shape. Collect a variety of oddly shaped empty containers—soda bottles, condiment dispensers, shampoo bottles, shoeboxes, etc. Compare the capacity of a long slender container with that of a short wide container by filling one container with water or sand and then pouring it into another container.

This is a good time to go over square and cubic units. Explain that when a unit is multiplied by itself, the answer is in squared units (ft times ft = ft^2, then ft^2 times ft = ft^3).

NOTES

Chapter 5

Breaking Ground

"Get your survey notes out," Mr. Whyte said. "We need to calculate the volume of cut and fill for site preparation at the Brown house."

"What about the homework, Mr. Whyte?" Phil asked. "I have a bunch of questions."

"Homework's on volume, so we can do it as we work through the calculations," Mr. Whyte said. "Al, what's volume?"

"Volume is the amount of material occupying a three-dimensional space."

"Sounds like you read the book. How about trying in your own words?"

Al gazed into space for a moment and then said, "It's something that takes up space and can be measured."

"Good. And how is it measured? Travis?"

"Cubic units, like inches, feet, and yards."

"Good. What's volume based on? Everyone?"

Faces broke into smiles and the class mumbled, "Surface area."

"I can't hear you!"

"Surface area!"

"Good. Glad you're all awake. So volume is the amount of space a material takes up and it has three dimensions—length, width, and depth." Mr. Whyte wrote on the board. "This is a rectangular solid, and the formula for finding its volume is length times width times depth or height. Since length times width is the formula for finding the area of a rectangle, the formula for the volume of a rectangular solid is area times depth."

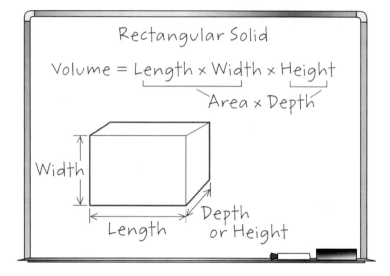

Key To Understanding

Volume is the amount of space a solid figure occupies. It has three dimensions: length, width, and height.

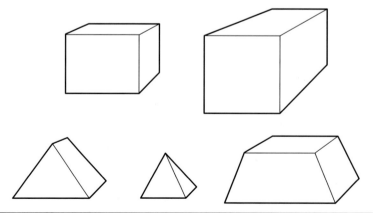

"And that same formula can be applied to a cube, except the length of the sides of a cube are equal, so the formula is side cubed or s to the third power. And since the formula for the area of a square is s^2, the formula for volume is area times the side."

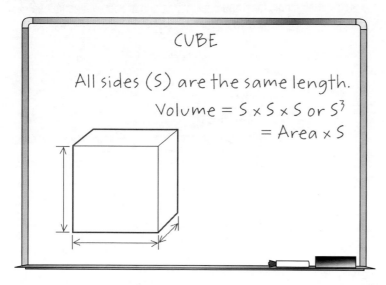

"Questions? None? Okay. First three homework problems. Travis, Al, and Sandy, up to the board. Al, you first."

"Mr. Whyte, I just realized I did the first problem wrong. I'm going to need to do it on the fly."

"What did you do wrong?"

"I multiplied inches times feet without converting first," Al said.

"Okay, go ahead. Tell us what you're doing as you go," Mr. Whyte said.

"The first question is for a rectangular solid." Al wrote on the board. "The length is 2 feet, the width is 6 inches, and the depth is 3 feet." Al wrote on the board.

 **Mind Games**

Area is measured in square feet because you are multiplying feet by feet, which is feet^2. Volume is measured in cubic feet because you are multiplying feet by feet by feet, which is feet^3.

When you need to divide two dissimilar units like dollars and hours, you need to combine the two units in the answer. For example, if you work 10 hours and you get paid $65, you calculate your hourly rate by dividing $65 by 10 hours:

$$\frac{\$65}{10 \text{ hours}} = \$6.50 \text{ per hour}$$

The answer is in the units of the numerator (top number) per the denominator (bottom number).

Here's another example: when 12,500 people live in an area of 5 square miles, you calculate the average number of people living in a square mile by dividing the total number of people by the area:

$$\frac{12,500 \text{ people}}{5 \text{ miles}^2}$$

$= 2,500$ people per mile^2 (also called a square mile)

 **Math Speak**

Volume is measured in cubic units, such as inches, feet, and yards. It is written as cubic inches, cubic feet, and cubic yards and is abbreviated inches^3, feet^3, and yards^3 or in^3, ft^3, and yds^3. Volume is based on surface area.

When a square is the base of a three-dimensional object, the object is called a cube.

Cube
Volume = S^3

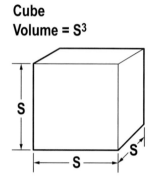

When a rectangle is the base of a three-dimensional object, it is called a rectangular object or sometimes a rectangular solid.

Rectangular object or solid
Volume = L × W × D

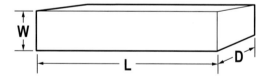

The volume of either object is calculated by multiplying its length by its width by its depth. Length is abbreviated with the letter L, the width with the letter W, and the depth with the letter D. Since all of the cube's dimensions are equal, each side can be abbreviated with the same letter (s), and the formula can be written s^3.

Key To Understanding

Convert all measurements to the same units before you use them to calculate area or volume.

- To convert feet to inches, multiply by 12.
- To convert inches to feet, divide by 12.
- To convert yards to feet, multiply by 3.
- To convert feet to yards, divide by 3.

"I didn't see that the width was in inches, so I should have converted it to feet. There are 12 inches in a foot so 6 inches equals 6/12 feet or ½ foot. Volume is length times width times depth, so the answer is 3 cubic feet."

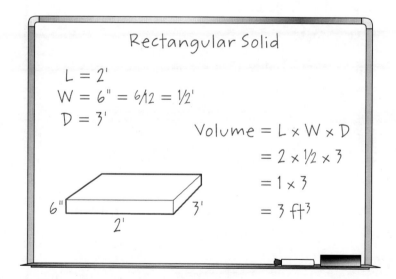

"Good job, Al." Mr. Whyte turned to the class. "Al brought up a good point. Before you perform any operations, convert all measurements to the same units."

Practice Problems 5-1

Answer the following questions. Draw the figure described in the question, if necessary.

1. Convert the following as indicated. Reduce all fractions.

   3 inches = **¼** feet

   15 feet = **5** yards

   144 inches = **12** feet = **4** yards

   1 foot 6 inches = **1½** feet

   54 inches = **4½** feet

2. Calculate the volume for a cube that is 2 feet in length, width, and depth.

   Volume = s³ = 2³ = 8 cubic feet

3. An object has a rectangular base with a surface area of 6 square feet and a depth of 11 feet. Calculate the volume of the object.

   Volume = 6 square feet × 11' = 66 cubic feet

4. Calculate the volume for the figure shown below. (Hint: convert inches to feet first.)

   Volume = 9' × ⅔' × 7' = 42 cubic feet

   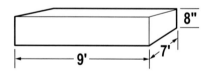

5. From the information given below, state the name of the object, and then calculate its volume.

   The shape is a cube because the sides are equal.

   S = 24" = ²⁴/₁₂ = 2'

   Volume = s³
   = 2' × 2' × 2'
   = 8 cubic feet

   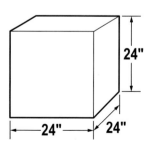

"Phil, did you get your questions answered yet?" Mr. Whyte asked.

"No, sir, my questions have to do with the next three problems."

"Yeah, I had problems with those too," Travis said.

"Me, too," Al and Sandy said together.

"Anybody able to work any of these questions out?" Mr. Whyte asked.

Jorge and Olivia raised their hands.

"Good. Up to the board and show us what you've got," Mr. Whyte said. "Olivia, you first."

"This shape is a triangular prism. The base and height are both 2 feet, and the depth is 4 feet, so the volume is 8 cubic feet," said Olivia.

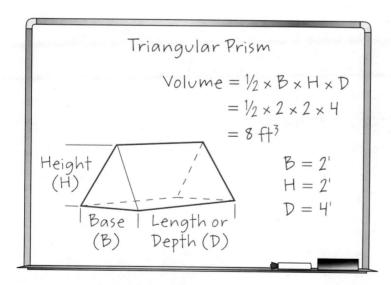

"Olivia, I don't think I know what a prism is, how did you decide it was a prism?" Phil asked.

"A prism is a shape that has two identical bases that are parallel to one another. See this front part is a triangle and the back part is a triangle." Olivia carefully colored in the two triangles.

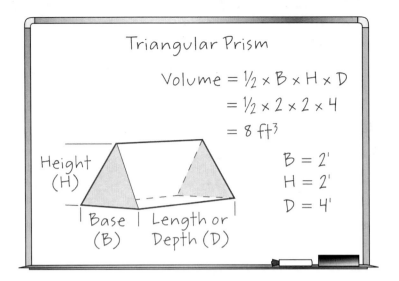

Key To Understanding

A prism is a multi-sided three-dimensional object that must meet all of the following requirements:

- It has two bases.
- The bases are parallel.
- The bases are the same shape.
- The bases are the same size.
- The remaining sides must be parallelograms.

All of the objects shown below are prisms. You will notice that rectangular objects and cubes are prisms, and so are triangles and trapezoids.

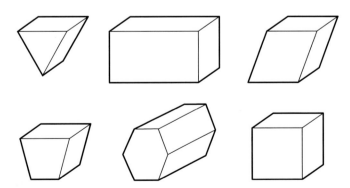

"Ohh. I see," Phil said.

"Me, too," Sandy said.

"Then I used the formula for the area of a triangle," Olivia continued. "And I multiplied that by the depth to get the volume."

"Good job, Olivia," Mr. Whyte said.

"Jorge, you're next," Mr. Whyte said.

"Well." Jorge smiled. "Mine's harder than Olivia's."

"Yeah, right," Olivia said.

"This shape is also a prism, but it has a trapezoid as its base. I found the area of the trapezoid and then multiplied it by the depth and got 4½ cubic feet for the volume."

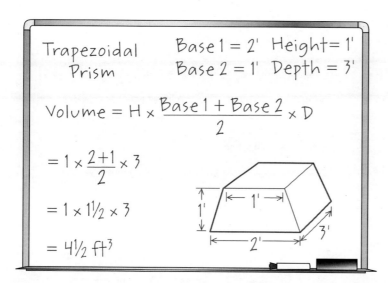

"Good. You guys can sit down." Mr. Whyte said, turning back to the class. "We need to talk about the volumes of other shapes, too, but we'll do it later. I want to get you guys going on calculating material for the Brown house."

 Math Speak

To calculate the volume of a prism, perform the following procedure:

Step 1 Identify the shape of the base.

Step 2 Identify the dimensions of the shape.

Step 3 Calculate the area of the base.

Step 4 Calculate the volume of the prism by multiplying the area by the depth.

For example:

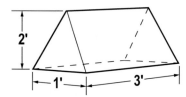

Step 1 The shape of the base is a triangle.

Step 2 The triangle's dimensions are Base = 1 foot and Height = 2 feet.

Step 3 The formula for a triangle is ½ Base times Height or ½BH, so the area is:

½ × 1 × 2 = 1 square foot

Step 4 If the area is 1 square foot and the depth is 3 feet, the volume is:

1 × 3 = 3 cubic feet

Practice Problems 5-2

1. Calculate the volume of a cube that is 2 feet in length, width, and depth.

   2^3 = 8 cubic feet

2. An object has a rectangular base with a surface area of 6 square feet and a depth of 11 feet. Calculate the volume of the object.

   6 square feet × 11 feet = 66 cubic feet

3. Calculate the volume of an object with a triangular base that measures 2 feet and a height of 6 feet. The object has a depth of 6 inches.

   2' × 6' × ½' = 6 cubic feet

4. An object has a base shape of a triangle with a base of 6 inches and a height of 1½ inches. The object is 3 inches deep. Calculate the volume of the object.

   ½ × (6" × 1½" × 3") = 13½ cubic inches

5. An object has a trapezoidal base with the dimensions of Base 1 = 5 feet, Base 2 = 11 feet, and Height = 2 feet. The object's depth is 25 feet. Calculate the volume.

   (5' + 11')/2 × 2' × 25' =

   ¹⁶⁄₂ × 2 × 25 =

   8 × 2 × 25 = 400 cubic feet

"Everyone remember how the Browns' property slopes?" Mr. Whyte continued.

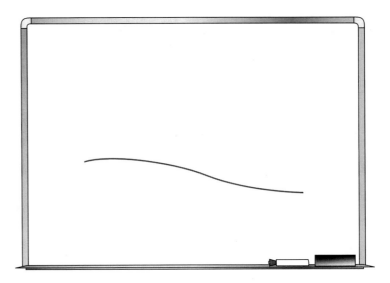

"How are we going to build a nice level house on that property?"

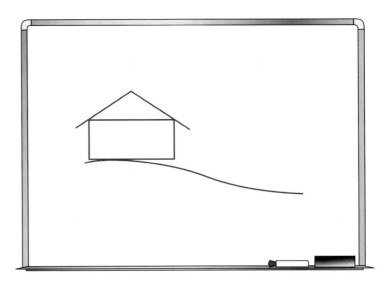

"We can fill it in with concrete," Travis said.

"We could, Travis. But concrete costs around $87 a cubic yard plus fuel and other delivery fees. That's a lot of money when we have something else available. Tom, what do you think?"

Tom was slouched in his chair staring blankly at his desk.

"Tom," Mr. Whyte repeated.

Phil gave Tom a tap on his shoulder to get his attention.

"Earth to Tom," Mr. Whyte said. "Do you copy?"

"What?" Tom swiveled his head first to Phil then to Mr. Whyte. "Um, excuse me. I didn't hear what you said, Mr. Whyte."

"You didn't hear because you weren't paying attention," Mr. Whyte said. "How are we going to level the Browns' property for the house?"

Color rose in Tom's cheeks. "Umm…we'll need a bulldozer or something to take the high spots down and then add fill dirt to the low spots."

"That's right," Mr. Whyte said. "What we'll do is shave off all of the grass and other vegetation—that stuff can't be used for fill dirt because plants decay and lose volume. What would happen if we put plants and tree trunks in the ground and then built something on it? Tom?"

Tom straightened in his chair. "As the plants and wood decay they lose volume, so the ground will settle and sink. Anything built on the ground would sink too, so that part of the foundation would crack."

"Excellent."

"Now, what were the house site elevations?" Mr. Whyte asked.

"The front of the house is 147 feet and the back is 146 feet," Jorge read from his notebook.

"The back part of the house is a foot lower than the front, so we need to cut a little soil from the front of the house and use it to build up the ground at the back of the house. If we need more fill, we'll have to order it

Key To Understanding

When calculating the volume of an object, all of the numbers must be in the same unit of measure—that is, all the numbers must be in inches, feet, or yards. It is a good idea to use the smallest measure you can. For example, 1 yard equals 3 feet, and 3 feet equal 36 inches, so the volume of a box that has a length, width, and depth of 1 yard can be calculated as follows:

Volume in yards:

Volume = 1 yd × 1 yd × 1 yd
= 1 yd^3

Volume in feet:

Volume = 3 × 3 × 3
= 27 ft^3

Volume in inches:

Volume = 36 × 36 × 36
= 46,656 in^3

from a vendor and pay for it. We are going to estimate the amount of fill we need. We want to get as close as we can, but we're not going to go crazy. So this line represents the ground."

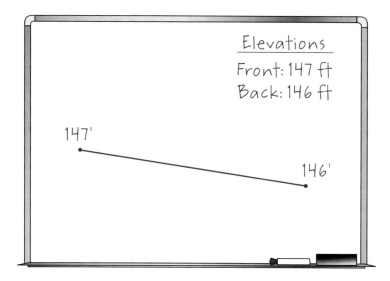

"Now, I know that the ground isn't a nice smooth **linear** descent. Some places are higher and some places are lower, but I'm estimating here, so I'll pretend that it is."

"Are we back to the fantasy world, Mr. Whyte?" Al asked.

"You bet, Al," Mr. Whyte said, glancing at Tom. Tom was back to slouching in his chair and staring at his desk. "Tom, how thick is the house slab?"

Tom looked up at Mr. Whyte and for a moment stared blankly at him. "It's …um… 4 inches, Mr. Whyte."

Mr. Whyte drew on the board. "Here's the slab superimposed on the ground. And look what I have—two triangles."

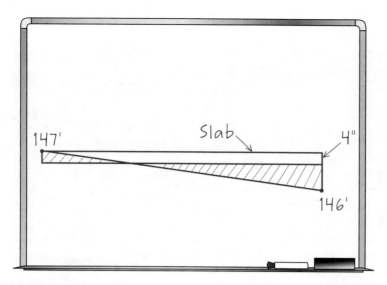

"The length of the house is 37 feet, and I know that up to here is 10 feet, because Mr. Burke told me." Mr. Whyte drew a line on the board. "So what's the length of the rest of it?"

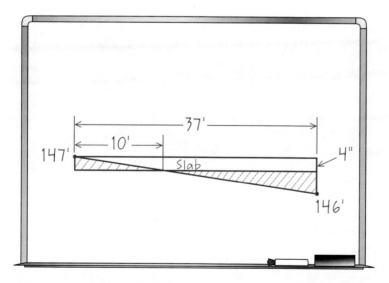

"If the whole length is 37 feet, and part of it is 10 feet, then the rest of it needs to be 27 feet," Al said.

"Good."

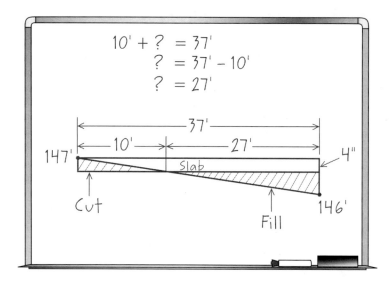

"Looking at the shaded areas, you can see there is a cut area and a fill area. What's the shape of both shaded areas?"

"Triangles, Mr. Whyte," Tom said. "Right triangles."

"Triangles." Mr. Whyte nodded. "And we need to find the volume of earth to cut from here." Mr. Whyte pointed to the small triangle. "And to fill here." He pointed to the large triangle. "What's volume based on?"

"Surface area!"

"I can't hear you."

"SURFACE AREA!"

"By Jove, I think you've got it. What's the formula for the area of a triangle? Olivia?"

"It's one half the base times the height."

"Good. Let's take a look at the small triangle."

Key To Understanding

Sometimes you need to look at the information you have and use it to calculate other information, just like Al did in the previous problem. For example, if you are working on a construction job and you need to move 100 cubic yards of dirt from a job site that has the dimensions of 30 feet by 90 feet, you can calculate the depth of the excavation, because volume is calculated by multiplying length, width, and depth:

Step 1 Convert feet to yards (3 feet equals 1 yard):

L = 30 feet = 30/3 = 10 yards

W = 90 feet = 90/3 = 30 yards

Depth = ?

Step 2 Fill in the information that you know:

L × W × D = Volume

10 yds × 30 yds × D = 100 yds^3

Step 3 Perform any calculation:

300 yds^2 × D = 100 yds^3

Step 4 Isolate the unknown:

$$D = \frac{100 \text{ yds}^3}{300 \text{ yds}^2}$$

Step 5 Solve for the unknown:

D = ⅓ yd

Step 6 Convert the answer to the original units:

D = ⅓ yd × 3

D = 1 foot

The depth of the excavation is 1 foot.

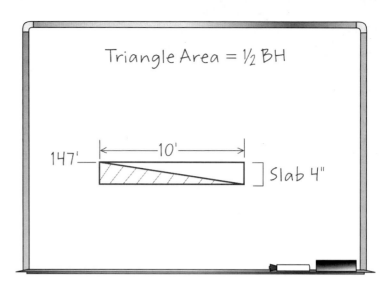

"What's the base measurement? Tom?"

"It's 10 feet."

"And the height?"

"4 inches."

"Everyone get that? The base is 10 feet, because that's the point where the slab **intersects** the ground. And the height is 4 inches, because that is the depth of the slab."

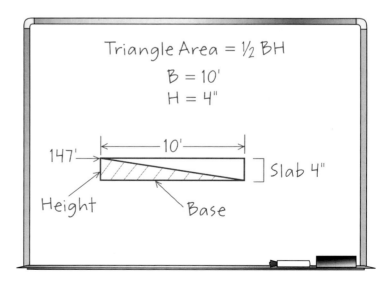

 **Mind Games**

When you look at a triangle, don't get stuck on its position. The following diagram shows the same triangle in different positions. See how different they can look!

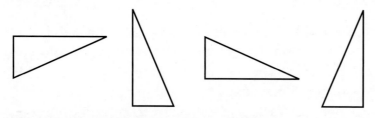

"We'll leave the problem there and go on to the large triangle. This one's a little harder, because we need to use some other information to calculate the base and height. Drawing just the big triangle, you can see that the base is 27 feet, but what about the height?"

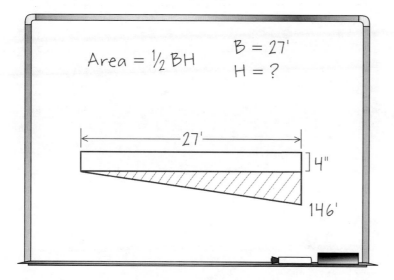

"Well," Travis said slowly. "We know the slab is 4 inches and the triangle starts—or ends, depending on how you look at it—at 146 feet, but I know that we can't just subtract 4 inches from 147 feet. That would be way too big."

"You're on the right track, Travis. Can anyone help him out?"

"Yeah," Sandy said. "The 146 feet is elevation, and the elevation at the front of the house is 147 feet. It's really only a foot from the top of the slab to the bottom of the triangle."

"Excellent, Sandy, excellent. The top of the slab is 147 feet in elevation. Remember what Mr. Burke told you about elevation. The lowest spot at the house site is 146 feet, so the difference is 147 feet minus 146 feet, or 1 foot. Then, as Travis said, we need to consider the slab, so the height of the triangle is 147 feet minus 146 feet and 4 inches. Okay, I'm going to stop here. For homework, I want you to calculate the volume of the cut and the fill. Remember to convert inches to feet and feet to inches as necessary."

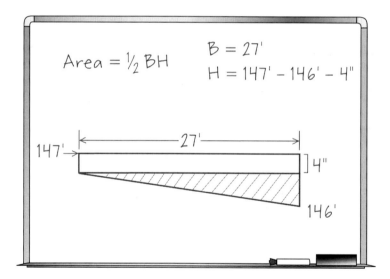

Mr. Whyte turned to erase the board.

"Um...Mr. Whyte," Al said. "Didn't you forget something?"

"What would that be, Al?"

"The width of the slab."

Mr. Whyte smiled. "Right you are, Al. It's 64 feet."

Mr. Provost came into the classroom and whispered something to Mr. Whyte. Mr. Whyte's eyes opened wide at whatever Mr. Provost told him and then he nodded his head.

"I need to leave. Start working on your homework," Mr. Whyte said as he headed for the door.

The room was silent for a minute after the door closed behind Mr. Whyte. Then Al turned to Tom and said, "Man, you are lucky Mr. Whyte got called away. He is going to have you on toast when he gets a hold of you. What's with you?"

Tom looked at Al and slowly shook his head.

Chapter 5 Review Problems

Calculate the volume of the following objects. (Hint: Draw the shape of the object if you need help visualizing it.)

1. Complete the classroom assignment for the cut and fill area of the Brown house site preparation.

   Cut = $\frac{4}{12}'$ × 10' × 64' = 213.33 cubic feet ÷ 27 = 7.9 cubic yards
   Fill = $\frac{8}{12}'$ × 27' × 64' = 1,152 cubic feet ÷ 27 = 42.67 cubic yards

2. You need to dig a hole that is 2 feet in length, width, and depth. The corners of the hole need to be 90 degrees. What is the shape of the object and its volume?

   Object is a cube; Volume = 2' × 2' × 2' = 8 feet

3. The dumpster at a work site has a rectangular base with a surface area of 24 square feet and a depth of 11 feet. What is the most the dumpster can hold without heaping the debris over the top?

   Volume = 24 square feet × 11 feet = 264 cubic feet

4. Mrs. Brown wants the class to make some concrete forms that she can use as plant stands in the yard. The forms have a triangular base that measures 2 feet with a height of 3 feet. She wants three forms: one is 24 inches tall, the second is 36 inches tall, and the third is 48 inches tall. What is the volume of each?

   Object 1 = ½ (2' × 3' × 2') = 6 cubic feet
   Object 2 = ½ (2' × 3' × 3') = 9 cubic feet
   Object 3 = ½ (2' × 3' × 4') = 12 cubic feet

5. Mr. Brown is thinking about putting a koi pond in the backyard. The pond has a trapezoidal base with the dimensions of Base 1 = 5 feet, Base 2 = 11 feet, and Height = 2 feet. It is 25 feet long. How much soil needs to be excavated?

   Volume = [(5' + 11') ÷ 2] × 2' × 25' = 400 cubic feet

6. There is an area of the Brown property that is prone to flooding in a heavy rain. The Browns got a permit to divert the water by digging a ditch that intersects with the one by the road. The ditch is trapezoidal in shape and has the dimensions of Base 1 = 3 feet, Base 2 = 7 feet, and Height = 3 feet. It is 9 feet long. How much soil needs to be excavated?

   Volume = [(3' + 7') ÷ 2] × 3' × 9' = 135 cubic feet

 **Careers in Construction— Heavy Equipment Operators**

 **Terms**

Intersect: To cross at a point.

Linear: Straight.

The construction industry always needs good heavy equipment operators. Heavy equipment operators work outside most of the time and need to be able to operate a variety of equipment, such as bulldozers, graders, backhoes, and dump trucks. Although entry-level operators don't need a lot of math skills, those who want to advance need to be able to calculate how long it will take to do a job and how much it will cost. This is especially important if you want to have your own business.

Some employers will teach a worker how to operate heavy equipment, but skills about estimating and how to manage a project probably need to be learned off the job. If you are disciplined enough to make yourself study, you can learn advanced skills by reading textbooks and equipment performance specifications. Those of you who want more formal instruction can attend a Heavy Equipment Operator's course. The National Center for Construction Education and Research (NCCER) accredits these types of courses, so see the NCCER web site (www.nccer.org) for more information and to find an accredited institution near you.

 **Did You Know**

When a property survey for residential construction is done, the surveyor doesn't need to know the exact elevation of the property with respect to sea level. The builder is mostly interested in making sure the house site is level. The builder can select any point as the reference and base cut and fill requirements on it.

On some construction jobs it's critical to know the exact elevation of the site with respect to sea level. In this case, the surveyor locates an official benchmark that has been installed by a government agency. Sometimes it's a local government agency, but often it's the National Geodetic Survey, which is a federal program. Benchmarks show the exact elevation of a point in relation to sea level.

Global Positioning Satellites (GPS) systems are used in surveys. The GPS program was developed by the U.S. military to provide precise position information for vehicle, troop, and equipment movement. It's made up of a series of satellites that circle the earth at various orbiting levels. These satellites broadcast a radio signal to earth that contains information about each satellite's location. A GPS surveying system uses an electronic receiver to receive these signals, and then a computer in the receiver calculates the exact position of the receiver. Since the GPS devices are radio receivers, they do not have the line-of-sight limitations of manual equipment—like the kind Mr. Burke showed the class—and laser instruments.

GPS systems are often used to determine distance and elevation, but while GPS devices are very reliable for distance, they can be unreliable for elevation. Since the GPS doesn't have a point of reference for elevation, it can't calculate the exact elevation. Some GPS systems use two receivers. One receiver is placed in a fixed location and becomes the reference. The other receiver can be moved to different positions, and computers will calculate the exact location (bearing and elevation) based on the reference.

NOTES

Chapter 6
Payday

In this chapter, students get paid for their work and need to learn about taxes. Taxes are in percentages. Since almost everyone can relate to money, demonstrate percentages by placing 100 pennies in a circle to make a pie chart. Explain that each penny represents 1% of the circle. Remove 5%, then 10%, then 50% from the pie. Next, line the 100 pennies up in a straight line. Explain that the IRS taxes incomes at 10%, 15%, 25%, 28%, 33%, and 35% based on taxable income. Take 10 pennies away from the line to represent 10%, then 15 for 15%, etc.

In advanced classes, you can explain that the tax structure is weighed so that only amounts over a certain income are taxed at the higher level. The chart below shows the schedule for a single taxpayer. You can get the complete tax schedule from the IRS website.

	If TAXABLE INCOME		Then the TAX is		
	Is Over	But Not Over	This Amount	Plus This %	Of the Excess Over
Single	$0	$7,550	$0.00	10%	$0
	$7,550	$30,650	$755.00	15%	$7,550
	$30,650	$74,200	$4,220.00	25%	$30,650
	$74,200	$154,800	$15,107.50	28%	$74,200
	$154,800	$336,550	$37,675.50	33%	$154,800
	$336,550	—	$97,653.00	35%	$336,550

Paying taxes is a hot button topic for everyone, so expect a lot of discussion and disagreement. You can point out to the class that when they start working, they will likely pay little or no taxes because they need to make more than the standard deduction, which was $5,150 for a single person in tax year 2006. Note that there is usually a lot of confusion between the personal exemption and the standard deduction. Each wage earner is permitted to take the standard deduction. The personal exemption is claimed for dependent individuals. The person taking the exemption must contribute more than 50% of the support of the dependent.

NOTES

Chapter 6

Payday

"Everybody sit down," Phil said, peeking out the classroom door. "Here he comes."

The students hurried to their desks and sat with their hands folded, watching the door. A few grinned widely but most were trying to suppress smiles.

Mr. Whyte pulled open the door and looked in. "You guys are so quiet I figured you weren't here."

He walked over to his desk and placed a large white envelope in the center. All eyes rested on the envelope. "What's going on, people? It's as silent as a tomb in here. You'd think you were expecting something."

"Come on, Mr. Whyte," Al said. "We know you got the bread in there."

"Bread? Al, I don't have any bread in there."

"Come on, Mr. Whyte. Pay out."

Mr. Whyte smiled. "Okay, guys, payday! Everyone's check is exactly the same, so there's no need to compare them. Before I hand these checks out, I need to explain about payroll taxes."

"Explain away, Mr. Whyte, as long as it's not a math lesson," Al said.

Mr. Whyte laughed. "I can turn anything into a math lesson, Al. And this is no exception."

"You know, Mr. Whyte, I was just saying that to Phil. I said, 'Phil, that Mr. Whyte, he can make anything into a math lesson. I bet when he gives his kids their allowance he makes 'em recite multiplication tables. I bet he makes the waitress calculate her tip when he goes out to dinner. I bet if he met Tiger Woods he'd quiz him about his **handicap.**'"

"Well, Al, I'll have you know I have met Tiger Woods and he already knows how to calculate his handicap." Mr. Whyte punctuated his statement with a couple of shakes of his finger.

"Well, I'm glad all of that education paid off."

"May I continue?"

"Yes, sir, you have the floor."

"Thank you. You're each paid $6.15 per hour. Since we had those two teachers' conference days when there was no school, you guys were able to work 43 hours over the two-week pay period. How do I figure out the total wages?"

"Hours worked times pay per hour," Sandy said.

"Good. And that gives me what?" Mr. Whyte asked as he wrote on the board.

Most of the class scribbled on paper. "I got $264.45."

"Me, too."

"Ditto."

"Boy, when it comes to dollars and cents, you guys are good at multiplying decimals."

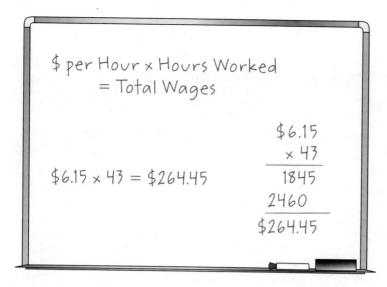

"Everyone's **gross** earnings are $264.45. Gross means before any deductions are made. In this country, almost all earnings are subjected to three types of **withholdings**."

"When you say withholdings, do you mean taxes?" Jorge asked.

"Yes, Jorge, in this country almost everyone needs to pay Social Security, Medicare, and income taxes. Your earnings after withholding are called your **net** earnings."

"I don't suppose a hardworking high school student who wants to buy a car to help his old grandmother with groceries is one of those people who don't pay taxes," Al said.

"You suppose right, Al," Mr. Whyte said. "Depending on how much you make in a year, you may not need to pay income taxes, but you will need to pay Social Security and Medicare taxes. These taxes are a **percent** of your gross earnings. What's percent, Travis?"

"Percent—that's part of a number."

"You're on the right track. Can anyone add to that?"

"It's a part of a number or thing—like a pie," Sandy said.

Math Speak

Percent is a part of 100. It is abbreviated with the symbol %. When you see a number and a percent sign like 10%, you say 10 percent.

"Percent is based on 100," Jorge said. "So when you say 10 percent, it means 10 pieces of 100."

"Good. Percent is part of 100." Mr. Whyte wrote on the board.

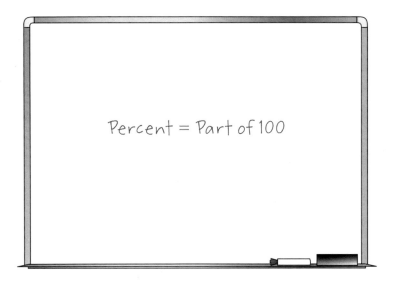

Percent = Part of 100

"So what's another name for part of something?"

"Fraction."

"Decimal."

Mr. Whyte nodded as he wrote on the board.

Percent = Part of 100
Percent is a fraction or decimal.

Key To Understanding

A percent is a fraction and its denominator is always 100, so any % equals the number/100.

$$10\% = \frac{10}{100} \quad 15\% = \frac{15}{100} \quad 5\% = \frac{5}{100}$$

Percent is also a decimal. Since percent is in 100s, the decimal equivalent of a whole is always at least to the second decimal place. You convert a percentage to a decimal by moving the decimal point to the left two places and dropping the percent sign, so the decimal equivalent of 15% is 0.15.

$$10\% = 0.10 \quad 15\% = 0.15 \quad 5\% = 0.05$$

A percentage can have a decimal part, too, but converting it to its decimal equivalent is the same—you move the decimal point to the left two places and drop the percent sign. The decimal equivalent of 15.5% is 0.155.

100% equals 1, and 150% equals 150/100 or 1.50.

Practice Problems 6-1

Convert the following percentages to their decimal equivalents.

1. 5% = **0.05**
2. 18% = **0.18**
3. 97% = **0.97**
4. 145% = **1.45**
5. 50% = **0.50**

Convert the following decimals to percentages.

6. 0.45 = **45%**
7. 0.88 = **88%**
8. 0.757 = **75.7%**
9. 1.37 = **137%**
10. 0.08 = **8%**

"Let's do some work with percent. If I am going to charge $125,000 to build a house and I'm going to spend 30% on labor, what's my total labor cost?" Mr. Whyte asked.

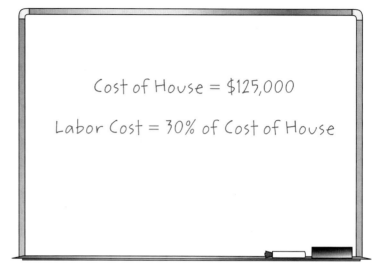

"First, how do I set up the problem? Jorge, what do you think?"

"I think that because $125,000 is the whole and 30% is the part, you need to multiply $125,000 by 30% to find out your labor cost."

"Very good, Jorge. Everyone get that?"

"No, Mr. Whyte, I don't get it," Al said.

"Me neither, "Olivia said. "What whole and what part?"

"Jorge, can you explain it?" Mr. Whyte asked.

"I can draw a picture of it."

Mr. Whyte handed the marker to Jorge. "You're on."

Jorge wrote on the board. "Remember how Sandy said that a percent is a part of something like a pie? This is the house pie. It equals 100% and is $125,000."

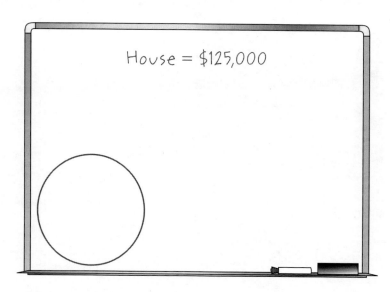

"Labor is a piece of the pie. It's 30%." Jorge drew on the board.

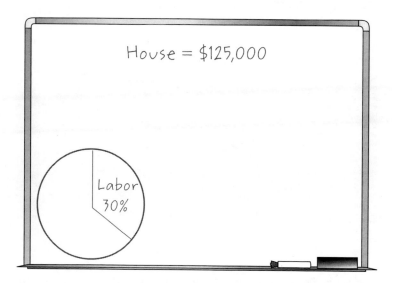

"To find the labor cost, you multiply its part of the pie in percent by the size of the pie."

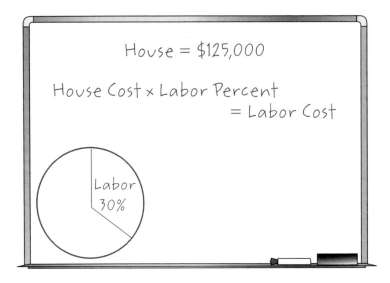

"$125,000 times 30% equals labor cost."

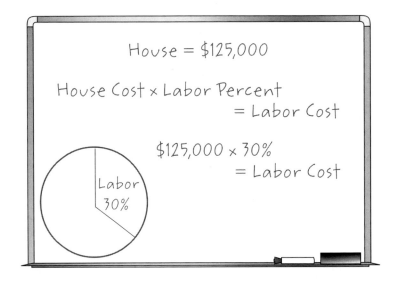

"So everything that costs money to build that house needs to come out of that pie?" Al asked.

"Exactly," Mr. Whyte said. "Thanks, Jorge, good job. In the house building business, you have labor and material costs, **overhead**, and profit." Mr. Whyte drew another circle on the board and then divided it. "The idea in business is to keep all of your costs as low as possible and your profit as high as possible."

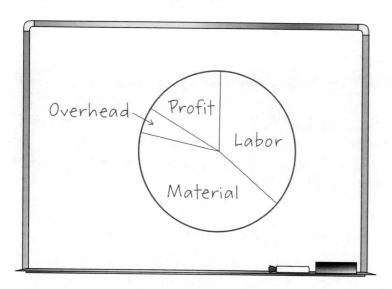

"What's overhead, Mr. Whyte," Jorge asked.

"It's costs that can't be directly attributed to building a house but you still need. Like office rent, your bookkeeper's salary, electricity in your office, or a copy machine."

Mr. Whyte pointed at Jorge's work on the board. "Back to Jorge's solution. We know that 0.30 is the decimal equivalent of 30%, so we need to multiply $125,000 by 0.30 to get our labor cost. And I get $37,500."

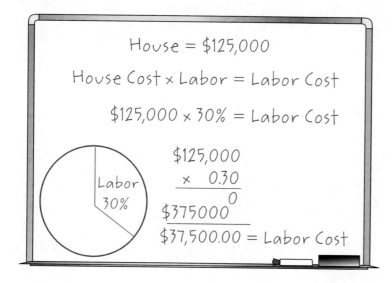

"Mr. Whyte, there are a lot of zeros there. Can we convert to powers of ten to make it easier?"

"Good catch, Al. Remember how when we used powers of 10 to calculate cost per square foot, we divided? When we divide powers of ten, we subtract the exponent in the divisor from the exponent in the dividend. In multiplication, you do the reverse—you add the exponents to get the answer."

"Nice to know, Mr. Whyte, but there's only one big number to convert to a power of 10. The $125,000."

"Ahh, my dear Al, we have a new wrinkle in the fabric of mathematics," Mr. Whyte said. "We have a small number—0.30 or 30%—and it can be converted to a negative power of ten. In the number 125,000, I move the decimal point to the left 3 times, so that's 10 to the positive third power. With 0.30, I move the decimal point to the right once, so that's 3 times 10 to the negative first power."

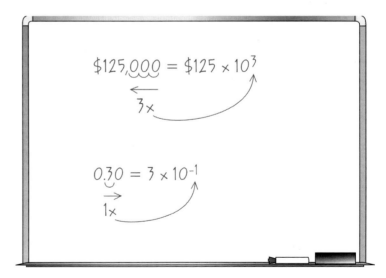

 Math Speak

When you divide with powers of 10, you subtract the exponent in the divisor from the exponent in the dividend to get the power of 10 for the quotient.

$$12 \times 10^2 \overline{)144 \times 10^3} \quad \begin{array}{c} 12 \times 10^1 \\ 3 - 2 = 1 \end{array}$$

When you multiply with powers of 10, you add the exponent in the multiplicand (the first number) to the exponent in the multiplier (the second number) to get the power of 10 for the product (the answer).

$$(12 \times 10^2) \times (12 \times 10^1) = 144 \times 10^3$$
$$2 + 1 = 3$$

 Practice Problems 6-2

Convert the following numbers as indicated.

1. $100 = 1 \times 10^2$
2. $0.330 = \underline{\mathbf{33}} \times 10^{-2}$
3. $1.25 = \underline{\mathbf{125}} \times 10^{-2}$
4. $0.097 = \underline{\mathbf{97}} \times 10^{-3}$
5. $3{,}250 = \underline{\mathbf{325}} \times 10^1$
6. $67 \times 10^{-3} = \underline{\mathbf{0.067}}$
7. $2 \times 10^{-7} = \underline{\mathbf{0.0000002}}$
8. $555 \times 10^{-9} = \underline{\mathbf{0.000000555}}$
9. $1 \times 10^{-1} = \underline{\mathbf{0.1}}$
10. $200 \times 10^{-4} = \underline{\mathbf{0.02}}$

Math Speak

When you have a decimal number, you can convert it to negative powers of 10 by moving the decimal point to the right and making the exponent negative.

$$0.03 = 3 \times 10^{-2} \qquad 0.1 = 1 \times 10^{-1}$$

(move decimal 2× for 10^{-2}; 1× for 10^{-1})

"Now I rewrite the problem, solve it, and I get 375 times 10 to the second power. Ten to the second power is 10^2 or 100, so when I finish up I get $37,500."

$$\$125 \times 10^3$$
$$\times \quad 3 \times 10^{-1}$$
$$\overline{375 \times 10^2 = 375 \times 100 = \$37,500}$$

"Hey, Mr. Whyte, it seems to me there's an easier way to do that," Tom said.

"If it involves a calculator, I don't want to hear it." The class let out a short burst of laughter.

"No, sir, I've accepted fantasy land. Can I show you?"

Mr. Whyte handed the marker to Tom. "I think that it would be easier to convert $125,000 to 1,250 times 10 to the positive second. And 30% to 30 times 10 to the negative second."

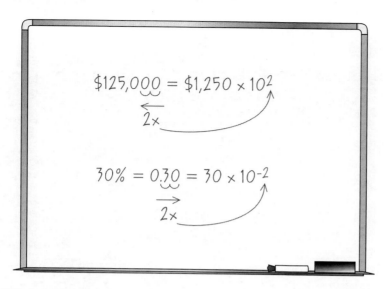

"Then when you multiply, you add the exponents and get 10 to the zero power, which is 1. See, no multiplication 'cause any number multiplied by 1 is itself."

"Good point. If it's easier for you to do it that way, do it. Let's get back to payroll withholding. First, there's Medicare tax. Just about everyone pays Medicare taxes. It's 1.45% of your gross earnings."

"Uh-oh. There's a decimal in that percent. If a percent is a decimal that means we've got decimals in decimals," Al said.

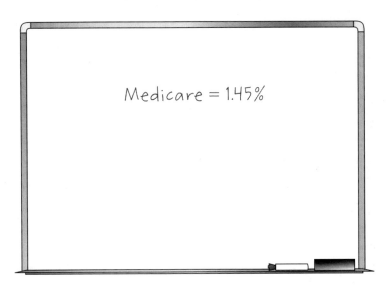

"Nothing we can't handle, Al. Then there's Social Security—it's 6.2% of your gross earnings. So all together we need to take 7.65% off of your earnings."

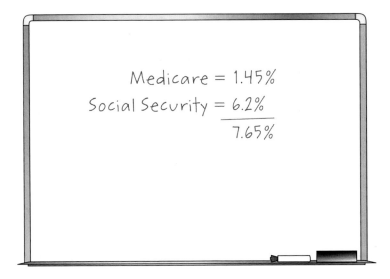

"Mr. Whyte, my grandmother gets Social Security checks. Does that tax have anything to do with those checks?" Al asked.

"Yes, Social Security taxes collected from payroll checks are used to pay the people who collect Social Security. Now, we have earnings of $264.45 and need to deduct 7.65% of it."

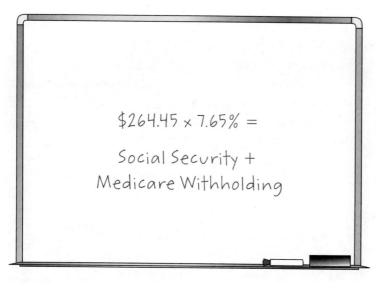

"Since we already know that any percent is in hundredths, we'll convert the percent to its decimal equivalent, which is 0.0765, and complete the problem.

"Move the decimal point in the answer 4 times for the 0.0765 and twice for the 264.45. Round to the second decimal, since we're talking about money, and I get $20.23 withholding for Social Security and Medicare."

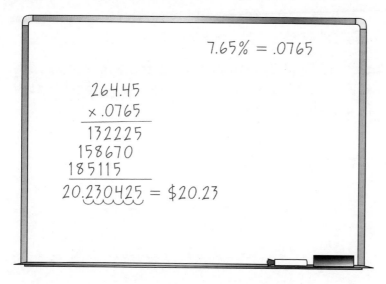

"Twenty bucks," Al said. "Man, that's a lot of money."

"We're not done yet, Al. We still need to deduct income tax," Mr. Whyte said. "It's a bit complicated to figure out income tax on salary, so I'm going to tell you that you each need to pay $3.55 in income tax on these earnings. With Social Security and Medicare that's a total of $23.78 from your paycheck that goes to Uncle Sam, leaving you with a check of $240.67."

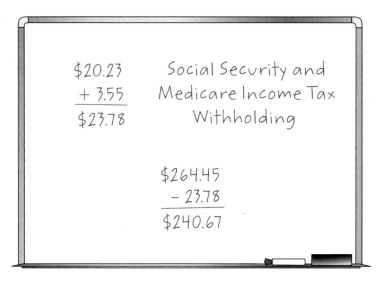

The class stared at the numbers on the board. After a minute, Olivia said, "That's a lot of money."

"And that's with no state income tax. Florida doesn't have income tax but most states do, so in those states more money would be withheld."

"Mr. Whyte," Jorge said. "What does the government do with the taxes?"

"A lot of it goes to Social Security, our country's defense—you know, the military—highways, education, paying interest on the national debt, stuff like that."

"I bet a lot of it goes to other countries," Jorge said.

"Actually, less than 1% of the U.S. budget for 2004 was for foreign aid," Mr. Whyte said.

"That's real interesting, Mr. Whyte, but what's even more interesting is this check plus what I saved working for my uncle last summer is enough for me to buy my car this weekend," Al said.

Chapter 6 Review Problems

1. Percent can be applied to any number. If the living room of a house is 20% of the 2,000 square foot house, what is the area of the living room?

   2,000 square feet × 20% =

   2,000 square feet × 0.20 = 400 square feet

2. You finally graduated from college and you are hired as an entry-level accountant at Stacey Construction, Inc. You make $2,000 every two weeks, and you need to pay 7.65% in Social Security and Medicare withholdings, and $283.09 in income tax. How much money do you get in your check?

   $2,000 × 7.65% =

   $2,000 × 0.0765 = $153.00

   $2,000 − $153.00 − $283.09 = $1,563.91

3. There are 25 students in Sandy's history class, and 16% failed the last test. How many students failed the test? How many passed?

   25 × 16% =

   25 × 0.16 = 4 failed, so 21 passed

4. Mr. Whyte calculated that the labor costs for building a house will be 28% of the selling price of $140,000. What is the total labor cost?

   $140,000 × 28% =

   $140,000 × 0.28 = $39,200

5. Mr. Whyte calculated that the expenses for building a house that has a selling price of $150,000 are as follows:

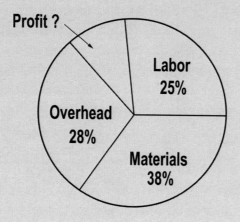

   a. Profit is <u>9</u>% of the total. **(25% + 38% + 28% = 91%, so 100% − 91% = 9%)**

   b. Profit is <u>$13,500</u>. **($150,000 × 9% = $13,500)**

   c. Labor costs are <u>$37,500</u>. **($150,000 × 25% = $37,500)**

   d. Material costs are <u>$57,000</u>. **($150,000 × 38% = $57,000)**

   e. Overhead costs are <u>$42,000</u>. **($150,000 × 28% = $42,000)**

6. Look at the overhead costs in Question 5. Overhead costs are those costs that can't be directly attributed to the building of the house—like rent for office space or electricity. List at least five other items that could be considered overhead.

   Answers will vary but can include: rent, utilities (examples include gas, electric, telephone, and Internet access), insurance on buildings and equipment, administrative staff (payroll, secretary, accounts receivable and payable, material clerk), equipment maintenance, oil and fuel costs, advertising and marketing costs, office furniture, office supplies, office equipment, property taxes, uniforms and laundry, personal protective equipment, etc.

 **Terms**

Gross: The amount of something without any deductions applied. Gross earnings are earnings before taxes and other withholdings are deducted.

Handicap: In golf, the average number of strokes of a golfer's score over the golf course's average score. A golfer's handicap is subtracted from his or her score.

Net: The amount of something after deductions are applied. Net earnings are earnings after taxes and other withholdings are deducted.

Overhead: Any business expenses not directly related to a project.

Percent: A portion of a number expressed in hundredths.

Withholding: An amount deducted from one's earnings.

Careers in Construction— Payroll Administrator

Payroll administrators may be accountants or accounting clerks. They are responsible for ensuring that a person is paid for the hours worked and that the proper amount is withheld from their earnings. They must also be sure that the money withheld is sent to the correct government agency at the correct time. People who handle payrolls must be very responsible and honest. They need to be very good at math, because people don't want mistakes in their paychecks. And just like people, the U.S. government wants to get withholding taxes quickly. Some payroll administrators have accounting degrees, but others learn on the job. If you want to handle payroll for a company, you need to be good at math, good with details, and very honest.

 Did You Know

Employees pay 1.45% of their wages for Medicare and 6.2% of their wages for Social Security, but employers need to pay the same amount per employee. So for every dollar that you pay into these two programs, your employer pays a dollar, too. That's in addition to the Social Security and Medicare that the employer pays on his or her own salary.

History—Medicare Program

In 1965, President Lyndon B. Johnson signed the Medicare bill. Medicare is a federally-funded health insurance program for people over the age of 65 or who have certain disabilities. This program is funded by the Medicare tax that workers pay.

History—Social Security Program

Social Security is a federal program that pays elderly or disabled workers a monthly income. Social Security also covers the dependent children of a deceased worker. Payments into Social Security are mandatory, and the amount of a person's monthly benefit is based on their lifetime earnings.

Social Security was signed into law in 1935 by President Franklin D. Roosevelt. This was after the Great Depression when a lot of people—especially the elderly—were very poor. Originally, it was just supposed to provide a small income after retirement, but over the years, it's been modified by Congress to include disabled people and dependents of a deceased worker.

Social Security benefits are small. The program was designed to give people a little extra income. It was never designed to be a full retirement plan. Workers need to save for their retirement.

The first monthly Social Security check was issued on January 31, 1940 to Ida May Fuller of Ludlow, Vermont. Miss Fuller was a legal secretary, who retired in November 1939. She paid Social Security taxes for only three years and the total she paid was $24.75. Her first check was for $22.54. Miss Fuller died when she was 100 years old, so she collected Social Security for 35 years and her total Social Security benefits were $22,888.92!

NOTES

Chapter 7
Divide and Conquer

This chapter continues to discuss volume. The problems in the chapter are more complicated than those in Chapter 5, so it's important that the students understand how to break complex problems into easy-to-handle pieces and to write equations to solve the problems. Most students want to break shapes into cubes or rectangles, but you should encourage them to use prisms, also. You can use modeling clay to demonstrate volume. Reshape the clay into different shapes. Start with a cylinder, then a pyramid, and then a sphere to show the students that the same volume can take on many appearances.

Now is a good time to review the order in which mathematical functions should be performed: actions in parentheses first, followed by multiplication and division, and then addition and subtraction (PMDAS). Be certain to include exponents in the multiplication step. Instruct students to always move from left to right when performing multiplication and division, and then addition and subtraction. For example, the following equation can result in a number of incorrect answers if the PMDAS order is not followed:

$$(3 + 3) \times 2 - 6 \div 3 + 1 = ?$$

Step 1	Parentheses:	$(3 + 3) \times 2 - 6 \div 3 + 1 = ?$
Step 2	Multiply and divide:	$6 \times 2 - 6 \div 3 + 1 = ?$
		$12 - 6 \div 3 + 1 = ?$
Step 3	Add and subtract:	$12 - 2 + 1 = ?$
		$10 + 1 = ?$
	Result	11

When none of the numbers are grouped within parentheses, the process is as follows:

$$3 + 3 \times 2 - 6 \div 3 + 1 = ?$$

Step 1 Multiply and divide: $3 + 3 \times 2 - 6 \div 3 + 1 = ?$

$$3 + 6 - 6 \div 3 + 1 = ?$$

Step 2 Add and subtract: $3 + 6 - 2 + 1 = ?$

$$9 - 2 + 1 = ?$$

$$7 + 1 = ?$$

Result   8

Many of the equations used to calculate surface area and volume will require multiplying a series of numbers. You can help your students to understand that since the operation is the same, the order in which they are performed is not important. This is very handy when fractions are involved. For example, compare the following equations:

$3 \times 4 \times 2 \times ½ = ?$	$3 \times 4 \times 2 \times ½ = ?$	$3 \times 4 \times 2 \times ½ = ?$	$3 \times 4 \times 2 \times ½ = ?$
$12 \times 2 \times ½ = ?$	$3 \times 4 \times 1 = ?$	$3 \times 8 \times ½ = ?$	$3 \times 2 \times 2 = ?$
$24 \times ½ = ?$	$3 \times 4 = ?$	$24 \times ½ = ?$	$6 \times 2 = ?$
Result 12	Result 12	Result 12	Result 12

Chapter 7

Divide and Conquer

"There's been a delay in getting the permit for the Browns' house, so we can't start work on the house this week," Mr. Whyte said.

The class let out a collective groan.

"Oh, man," Al said, running a hand over his bald head. "I need to make $120 fast. That guy's not going to hold the Mustang for me much longer."

"What do you mean hold it, Al?" Phil asked. "After we got paid last week, you said you had enough money to buy the car."

Al slouched deeper in his chair. "I did. I had 80 bucks extra. It was my Gran's birthday on Saturday. She always wanted to go to that fancy restaurant in Gainesville. You know the one near the university where all the waiters dress in black and carry white towels on their arms. I figured 80 bucks would cover it, but the whole bill came to almost $200."

"Whoa, $200 for dinner? What were you guys eating?" Travis asked.

A smile twitched at Mr. Whyte's mouth. "I'm sure your Gran was pleased to get out to such a nice place."

"Yeah, she was. The food was really good, and everyone was real nice to Gran. They brought her a cake and everyone sang Happy Birthday to her. She started crying she was so happy." Al smiled a little at the memory.

"I have good news for you, Al," Mr. Whyte said. "Mr. Brown also wants to have a big shop built on the property, and he has the permit. The building's already contracted, but he needs a slab. He said we could do it."

"Oh, man. That's great, Mr. Whyte." Al sat straighter in his chair and smiled.

"Yeah, cool," Travis said.

"The shop's a steel building. I have the plans right here," Mr. Whyte said as he unfolded some large drawings and placed them on his big worktable. "Everyone gather 'round. The slab is easy. It's a monolithic slab. Jorge, what's a monolithic slab?"

"It's when the footings and the slab are poured together," Jorge said.

"Exactly. We need to get the building site prepared for a slab. This includes placing anchor bolts at the proper points," Mr. Whyte said.

"Mr. Whyte, I don't get this. What're the anchor bolts for?" Sandy asked.

"It's like this." Mr. Whyte turned to the whiteboard and drew a slab and a building. "We pour a slab and want to build a structure on it. In this case, it's a steel building, and we need something to hold it onto the slab."

Mr. Whyte continued. "To attach the building to the slab, we set anchor bolts so when the concrete for the slab is poured, the anchor bolts stick out of the slab. There are all sorts of anchor bolts—some are straight and others are curved, but they all stick out of the slab. That's where the support beams for the building are attached." Mr. Whyte pulled out a reference book and show the students a picture.

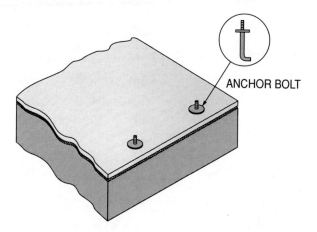

ANCHOR BOLT

"When we build the house," Mr. Whyte continued. "The wooden **sill plates** will be attached to the anchor bolts so we'll have some wood to nail the frame to."

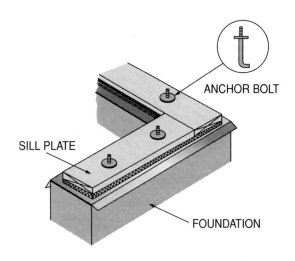

"Pretty neat, Mr. Whyte," Sandy said.

"I think so, too," Mr. Whyte told Sandy with a smile. "Okay, guys. Let's figure out how much concrete we need for the foundation. Tom, what do we need to know to figure out how much concrete we need for the slab?"

"Concrete is supplied in cubic yards, which is volume, so we need to calculate the volume of the slab," Tom replied.

"Good. And what's volume based on, Tom?"

Tom scratched his head and said, "Wait a minute, Mr. Whyte. It's coming to me." Tom smiled, squinted his eyes, and gazed off at the corner of the ceiling. He turned to Phil. "I just can't think of it. Do you remember what it is, Phil?"

Phil scratched his head. "I can't think of it either. What about you, Sandy?"

Sandy frowned. "It's on the tip of my tongue, but I can't get it out." She turned to Travis. "What is it, Travis?"

Mr. Whyte laughed. "Okay, guys, you've had your fun. Everyone, what's volume based on?"

"SURFACE AREA!"

"I'm glad that's settled. This building is 30 feet by 40 feet with a 6-inch slab. Tom, show us how to figure out the volume." Mr. Whyte handed Tom a marker.

"The formula for volume is surface area times depth. Since the building is 30 feet by 40 feet, the surface area is 1,200 square feet," Tom said as he scribbled on the whiteboard. "The slab is 6 inches thick. I need to convert the 6 inches into feet. Since a foot equals 12 inches, I put 6 over 12 and reduce the fraction to ½. That's ½ foot. Next, I multiply ½ by the area—1,200 square feet—and I get a volume of 600 cubic feet."

$$\text{Volume} = \text{Area} \times \text{Depth}$$
$$\text{Area} = 30' \times 40' = 1{,}200 \text{ ft}^2$$

The slab is 6" thick (depth).

$$6" = 6/12 = \tfrac{1}{2}' \text{ (convert inches to feet)}$$

$$\text{Volume} = 1{,}200 \times \tfrac{1}{2} = 600 \text{ ft}^3$$

"Good. Everyone write that in your notebook: Slab volume equals 600 cubic feet."

Practice Problems 7-1

Complete the following problems. Draw a diagram if necessary. Show all of your work. Since the answers are for volume of concrete, you can round up to the nearest whole number.

1. You are working on a house that will be built **slab-on-grade**. The foundation is a rectangle that is 60 feet long and 50 feet wide. The slab depth is 4 inches. What is the volume of the slab?

    Volume = 60' × 50' × 4/12' = 1,000 ft³

2. The class has been contracted to build an outdoor racquetball court. It is 20 feet wide and 40 feet long. The slab thickness is 4 inches. What is the volume of the slab?

    Volume = 20' × 40' × 4/12' = 267 ft³

3. Mrs. Brown is thinking of having a picnic pavilion built behind the house. She wants it to be 10 feet by 15 feet. Mr. Brown thinks the slab should be 3 inches thick, but Mr. Whyte thinks it should be at least 4 inches. What are the volumes of both slabs?

    Volume for 3" Slab   **Volume for 4" Slab**
    10' × 15' × 3/12' = 38 ft³   10' × 15' × 4/12' = 50 ft³

4. Mr. Brown is thinking of having an 8-inch slab for the pavilion in case he wants to park his truck there. What is the volume for this slab?

    Volume = 10' × 15' × 8/12' = 100 ft³

Mr. Whyte drew on the board. "This diagram represents the slab, and see the dotted line? That's the foundation footings." Mr. Whyte drew another diagram on the board. "And this is the footing detail."

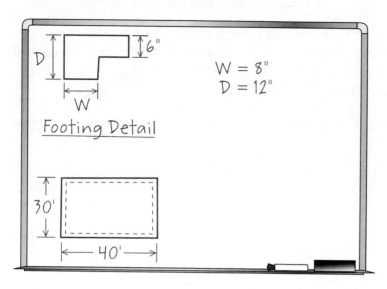

"Travis," Mr. Whyte asked. "What's the purpose of footings?"

"Footings carry the weight of the building's outside walls," Travis answered.

"Good. Footings are thicker than the rest of the slab. In cold climates, footings need to be dug down below the **frostline**. In this part of Florida, we don't get much cold weather so we can get away with shallow footings. Footings run around the edge of the foundation where the outside walls go. Since footings are deeper than the slab, we need to figure out the volume of the footings separately and add it to the slab volume. The details about the footings will appear on the foundation plans—like this." Mr. Whyte pointed to the footing detail on the whiteboard.

"If I wanted to calculate the volume of this footing, what do I do?" Mr. Whyte asked.

"Well," Sandy began. "You need length, width, and depth. I see the width—it's 8 inches. The depth is 12 inches, but what about the length?"

"Good question. Phil, can you help Sandy out?"

"I think you need to divide the footings up into pieces—kind of like we do when we need to calculate the area of an odd-shaped house," Phil said. "That footing at the bottom of the drawing is 40 feet long."

Mr. Whyte shaded the footing at the bottom of the drawing.

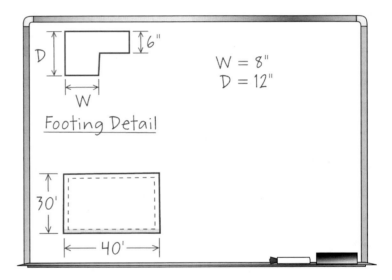

"Good. Does everyone see what Phil is saying? This footing is 40 feet long, 8 inches wide, and 12 inches deep," Mr. Whyte said. "Before we move on, I want everyone to look at the footing detail carefully. What can you tell me about the depth?"

The class stared at the diagram on the board. Finally, Olivia said, "It's the slab. The depth of the footing is 12 inches from the top of the slab, but we already calculated the volume of the slab, so we shouldn't include the slab depth in the footing depth."

"Very good, Olivia." Mr. Whyte shaded the top edge of the footing detail diagram. "The formula for volume is length times width times depth. The bottom footing is 40 feet long and 8 inches wide, and the depth is 12 inches minus 6 inches for the slab."

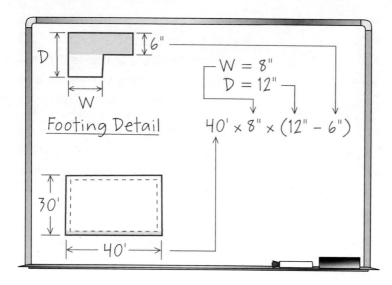

"Wait a minute, Mr. Whyte," Al said. "Why didn't we figure out the footing volume first so we wouldn't have to subtract the slab depth from the footing depth? It seems like this is an extra step."

"We could have done it that way, Al, but think about what you just said. If we used the full depth of the footing, what do we need to do to the dimensions of the slab?"

Al stared at the board. "Oh, I see. We would need to cut 8 inches off of each side of the slab, so the length would be...let's see." Al scribbled in his notebook. "The footing is 8 inches in width and there's one on each side so I need to subtract 16 inches from 40 feet." Al looked up at Mr. Whyte. "You know how much I love converting feet to inches and inches to feet. Sixteen inches is 1 foot 4 inches, so the length of the slab would be 38 feet 8 inches."

Math Speak

Remember, when you are multiplying fractions, you multiply numerators by numerators (top numbers) and denominators by denominators (bottom numbers). Then reduce the answer to its lowest term if necessary.

$$\frac{1}{2} \times \frac{14}{45} = \frac{14}{90} = \frac{7}{45}$$

"You got it. If you want to use the full thickness of the footing, you need to subtract the width of the footing from all four sides of the slab," Mr. Whyte said. "Okay, so what's the volume of the shaded footing? Sandy, up to the board and work it out for us."

Sandy went to the board and began writing. "Volume equals length times width times depth. Length is 40 feet, width is 8 inches, and the depth is 6 inches. I need to convert the inches to feet, so 8 inches is $8/12$ of a foot and 6 inches is $6/12$ of a foot." She paused for a minute.

"I can reduce both fractions," she continued. "The $8/12$ reduces to $2/3$, and the $6/12$ reduces to $1/2$. Now, I multiply 40 times $2/3$, which equals $80/3$. $80/3$ times $1/2$ equals $80/6$. That reduces to $13\ 2/6$. $2/6$ reduces to $1/3$, so the volume of this footing is $13\ 1/3$ cubic feet."

$$\begin{aligned}
\text{Volume} &= L \times W \times D \\
&= 40' \times 8" \times 6" \\
&= 40' \times {}^8\!/_{12}{}' \times {}^6\!/_{12}{}' \\
&= 40' \times {}^2\!/_3{}' \times {}^1\!/_2{}' \\
&= {}^{80}\!/_3 \times {}^1\!/_2 \\
&= {}^{80}\!/_6 \\
&= 13\,{}^2\!/_6 = 13\,{}^1\!/_3\ ft^3
\end{aligned}$$

$$\begin{array}{r} 13\ 2/6 \\ 6\overline{)80} \\ \underline{60} \\ 20 \\ \underline{18} \\ 2 \end{array}$$

"I think she's got it!" Al exclaimed.

Sandy grinned at Al.

"Good job, Sandy," Mr. Whyte said as he drew a diagram of the slab on the board. "Let's try to make it easy for us. The bottom edge footing is $13\ 1/3$ cubic feet in volume. What about the footing on the top edge?"

"It has to be the same," Sandy said.

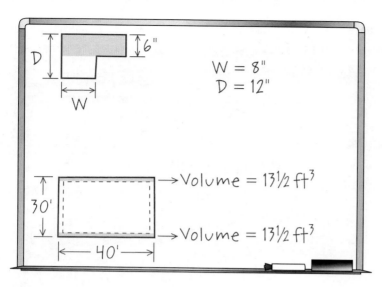

"Exactly. What about the 30-foot edge?" Mr. Whyte asked.

"It's just like Al said," Sandy answered. "We need to take 8 inches off the top and bottom. So when you want to figure out the volume for the 30-foot side, you need to use 28 feet 8 inches for the length. The other footing dimensions are the same—8 inches for the width, and 6 inches for the depth."

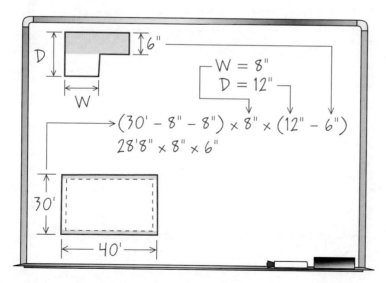

Mr. Whyte nodded. "Travis, show us how to calculate the volume of this footing."

Travis took the marker from Sandy and scribbled on the board. First, Travis converted the inches to feet and reduced them.

$$V = L \times W \times D$$
$$= 28'8" \times 8" \times 6"$$
$$= 28\tfrac{8}{12} \times \tfrac{8}{12} \times \tfrac{6}{12}$$
$$= 28\tfrac{2}{3} \times \tfrac{2}{3} \times \tfrac{1}{2}$$

Travis didn't want to convert the 28 feet to thirds, so he applied the commutative property and multiplied the 28⅔ by ½ and got 14⅓.

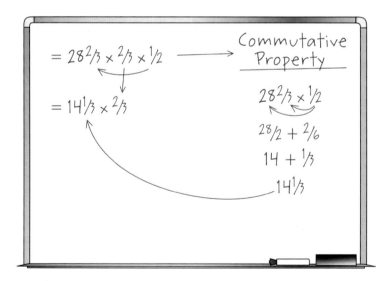

Then he finished the problem.

"I get 9⅗ cubic feet for the side footings," Travis said.

Key To Understanding

The commutative property of multiplication says that since the operation is the same, the order in which you perform it is not important. For example, compare the following equations:

3 × 4 × 2 × ½ = ?	3 × 4 × 2 × ½ = ?	3 × 4 × 2 × ½ = ?
12 × 2 × ½ = ?	3 × 4 × 1 = ?	3 × 8 × ½ = ?
24 × ½ = ?	3 × 4 = ?	24 × ½ = ?
Result 12	Result 12	Result 12

$$= 14\tfrac{1}{3} \times \tfrac{2}{3}$$

$$= \tfrac{43}{3} \times \tfrac{2}{3} = \tfrac{86}{9} \longrightarrow 9\overline{)86} = 9\tfrac{5}{9}$$

$$\phantom{= \tfrac{43}{3} \times \tfrac{2}{3} = \tfrac{86}{9} \longrightarrow} \underline{81}$$

$$\phantom{= \tfrac{43}{3} \times \tfrac{2}{3} = \tfrac{86}{9} \longrightarrow \ \ } 5$$

$$= 9\tfrac{5}{9} \text{ ft}^3$$

"Good job, Travis. We now have all we need to calculate the total volume of concrete we need for this foundation." Mr. Whyte wrote on the board. "The slab volume is 600 cubic feet, then we have two footings that are 13⅓ cubic feet and two more that are 9⅝ cubic feet. We do the multiplication first and reduce as necessary. To add these fractions, I need them to have the same denominator. Tom, what's the least common denominator for ⅔ and ½?"

> Volume =
> $= 600 \text{ ft}^3 + (2 \times 13\frac{1}{3} \text{ ft}^3) + (2 \times 9\frac{5}{9} \text{ ft}^3)$
> $= 600 \text{ ft}^3 + 26\frac{2}{3} \text{ ft}^3 + 18\frac{10}{9} \text{ ft}^3$
> $= 600 \text{ ft}^3 + 26\frac{2}{3} \text{ ft}^3 + 19\frac{1}{9} \text{ ft}^3$

"Um, 9, Mr. Whyte. Change that ⅔ to 6/9."

"Good," Mr. Whyte said, finishing the problem. "The total volume of concrete needed for this foundation is 645⅞ cubic feet. We'll round up to 646 cubic feet for simplicity."

> Volume =
> $= 600 \text{ ft}^3 + (2 \times 13\frac{1}{3} \text{ ft}^3) + (2 \times 9\frac{5}{9} \text{ ft}^3)$
> $= 600 \text{ ft}^3 + 26\frac{2}{3} \text{ ft}^3 + 18\frac{10}{9} \text{ ft}^3$
> $= 600 \text{ ft}^3 + 26\frac{2}{3} \text{ ft}^3 + 19\frac{1}{9} \text{ ft}^3$
> $= 600 \text{ ft}^3 + 26\frac{6}{9} \text{ ft}^3 + 19\frac{1}{9} \text{ ft}^3$
> $= 600 \text{ ft}^3 + 45\frac{7}{9} \text{ ft}^3 \ (45.775)$
> $= 645\frac{7}{9} \text{ ft}^3 \longrightarrow \text{round up to } 646 \text{ ft}^3$

Mr. Whyte tapped on the board with his marker. "Keep in mind that you could have set up this problem like this right from the start, but you would need to define some variables. Phil, what's a variable?"

"That's when a number is unknown, so you use a letter or symbol to represent it in an equation."

"Exactly." Mr. Whyte drew a simple slab on the board. "Let's name some variables. There are no dimensions on this diagram, but I still want to define its total volume, so I'll use variables. Tom, name a variable on this diagram."

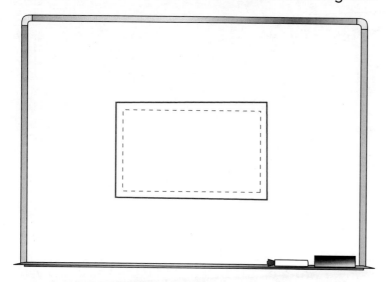

"Well, the slab is 30 feet by 40 feet, so that's not a variable...," Tom began.

"Nope. Pay attention. You're working on the last foundation. This one doesn't have any dimensions on it." Mr. Whyte pointed to the diagram on the board.

"Oh, yeah, so the slab volume is a variable."

"Right. Let's use the letter S as the slab volume variable. How many footings, Olivia?"

"Four. One on each side."

"That's true, but what can you tell me about the sides? Jorge?"

Jorge studied the diagram. "There are four sides—two lengths and two widths—but since the lengths are equal and the widths are equal, we can figure out the footing volume for one side and then multiply by 2."

 **Math Speak**

A variable is used as a placeholder for an unknown number. You use a variable just like a number.

An equation is a mathematical sentence or expression where both sides are equal. An equation defines something. In the following equation, you are defining volume:

$$\text{Volume} = \text{length} \times \text{width} \times \text{depth}$$

In this equation the length, width, and depth are the variables, because they are holding places for numbers. The variables can change, but the equation for volume is always the same. In the following equation, the length, width, and depth are defined with letters:

$$\text{Volume} = L \times W \times D$$

This equation can also be written without the multiplication symbols (Volume = LWD) because one rule of mathematics is when there is no mathematical symbol shown, the function is assumed to be multiplication.

"Precisely. Let's call the footing on the length side F_1 for Footing 1 and the one on the width side F_2 for Footing 2. You can call them anything you want, but you need to define them so everyone knows what you mean. So our equation for total volume is S_{vol} plus $2F_{1\,vol}$ plus $2F_{2\,vol}$."

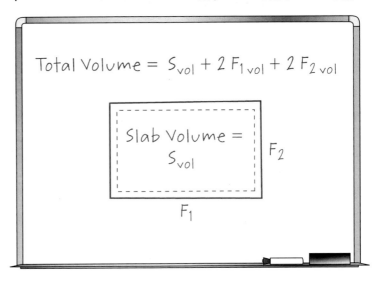

Mr. Whyte wrote on the board. "If I define all of my variables like this, what's the total volume?"

$$S_{vol} = 600 \text{ ft}^3$$
$$F_{1\,vol} = 20 \text{ ft}^3$$
$$F_{2\,vol} = 15 \text{ ft}^3$$

For a minute, the only sound in the room was the scratching of pencils on paper while the class calculated the volume.

"Show us what you've got, Travis." Mr. Whyte held a marker out.

"Well," Travis said, as he walked to the board and took the marker. "First, I substituted the numbers for the variables. I did the multiplication first, and then the addition. I got 670 cubic feet for the total volume."

$$\text{Total Volume} = S_{vol} + 2\,F_{1\,vol} + 2\,F_{2\,vol}$$
$$= 600 + 2 \times 20 + 2 \times 15$$
$$= 600 + 40 + 30$$
$$= 670 \text{ ft}^3$$

"Good, Travis. Questions?"

"Mr. Whyte." Jorge waved his hand in the air. "Didn't you tell us that concrete is ordered in cubic yards?"

Mr. Whyte twisted the end of his mustache and smiled. The class let out a collective groan. They knew Mr. Whyte would make them convert cubic feet to cubic yards.

"It's okay, everyone. Don't panic," Al said, walking to the board. "I can do this. A cubic yard is a block that's 1 yard by 1 yard by 1 yard. Since a yard is equal to 3 feet, a cubic yard is equal to 3 times 3 times 3 cubic feet—that's 27 cubic feet.

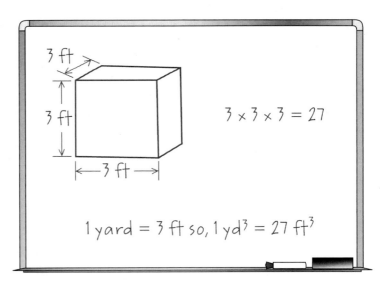

"To convert from cubic feet to cubic yards, I need to divide the cubic feet—which in this case is 670—by 27, because that's how many cubic feet are in a cubic yard. And I get 24.8…"

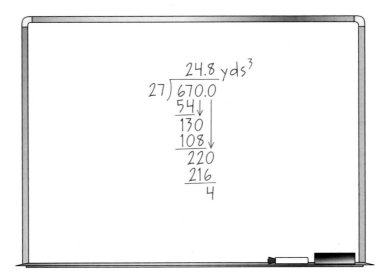

Mr. Whyte laughed. "Good job, Al. We'd round that number up and order at least 25 cubic yards of concrete. Maybe even more, because it's always better to be a little bit over when you're working with concrete. Concrete is now $87 a cubic yard, so how much will it cost for the slab?"

 **Mind Games** — Equations can be written for any problem. Read the following example carefully and then write an equation to solve for the slab and footing volumes.

The volume of the footing running the width of the foundation is 11 cubic feet. The volume of the footing running the length of the foundation is 5 times the volume of the width. The slab volume is 10 times the sum of the total footing volume. What is the total volume of the slab and footings?

Hint: Start with this equation: Total volume = $S_{vol} + 2F_{1\,vol} + 2F_{2\,vol}$

First, draw a diagram.

Next, label what you know:

a) The footing volume running the width of the foundation is 11 cubic feet.

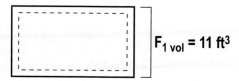

b) The footing volume running the length of the foundation is 5 times the volume of the width.

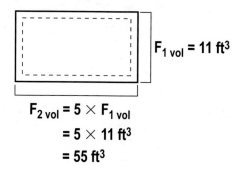

$F_{2\,vol} = 5 \times F_{1\,vol}$
$= 5 \times 11 \text{ ft}^3$
$= 55 \text{ ft}^3$

c) The slab volume is 10 times the sum of the total footing volume. (Hint: There are four footings.)

Slab Volume

$S_{vol} = 10 \times (2 \times 11 \text{ ft}^3 + 2 \times 55 \text{ ft}^3)$

$= 10 \times (22 \text{ ft}^3 + 110 \text{ ft}^3)$

$= 10 \times (132 \text{ ft}^3)$

$= 1,320 \text{ ft}^3$

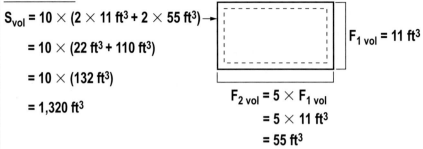

$F_{1\,vol} = 11 \text{ ft}^3$

$F_{2\,vol} = 5 \times F_{1\,vol}$

$= 5 \times 11 \text{ ft}^3$

$= 55 \text{ ft}^3$

d) Finally, calculate the total volume of the slab and footings:

Total Volume $= S_{vol} + 2F_{1\,vol} + 2F_{2\,vol}$

$= 1,320 + (2 \times 11) + (2 \times 55)$

$= 1,320 + 22 + 110$

$= 1,452 \text{ ft}^3$

e) Convert 1,452 cubic feet to cubic yards (1,452 ÷ 27 = 53.7778). Since this is concrete, round up to 54 cubic yards.

 Practice Problem 7-2

Complete the following problem. Be sure to convert cubic feet to cubic yards and round up to the nearest whole number when you calculate concrete costs.

1. Calculate the cost of the concrete for Mr. Brown's shop. Use a cost of $87/cubic yard for the concrete and a total volume of 25 cubic feet.

   Cost = 25 × $87

   Cost = $2,175

For Questions 2 and 3, calculate the total volume of concrete needed and then calculate its cost using $87/cubic yard. S_{vol} is the slab volume, $F_{1\,vol}$ is the volume for one footing on the length of the slab, and $F_{2\,vol}$ is the volume for one footing on the width of the slab.

2. S_{vol} = 750 cubic feet
   $F_{1\,vol}$ = 24⅝ cubic feet
   $F_{2\,vol}$ = 18⅜ cubic feet

   Total Volume = 750 ft³ + (2 × 24⅝ ft³) + (2 × 18⅜ ft³) = 836 ft³
   Convert ft³ to yd³ = 836 ft³ ÷ 27 ft³/yd³ = 31 yd³ (rounded up)
   Cost = 31 yd³ × $87 = $2,697

3. S_{vol} = 1,000 cubic feet
   $F_{1\,vol}$ = 50 cubic feet
   $F_{2\,vol}$ = 40 cubic feet

   Total Volume = 1,000 ft³ + (2 × 50 ft³) + (2 × 40 ft³) = 1,180 ft³
   Convert ft³ to yd³ = 1,180 ft³ ÷ 27 ft³/yd³ = 44 yd³ (rounded up)
   Cost = 44 yd³ × $87 = $3,828

4. The total volume of the slab and footings for a foundation is 800 cubic feet. The volume of the footings (less the slab thickness) are as shown below. $F_{1\,vol}$ is the volume for one footing on the length of the slab, and $F_{2\,vol}$ is the volume for one footing on the width of the slab. What is the volume of the slab? What is the total cost for the concrete at $87/cubic yard? Hint: Read this question carefully. You need to calculate the slab volume and then the cost for the total volume of concrete.

   $F_{1\,vol}$ = 30 cubic feet

   $F_{2\,vol}$ = 20 cubic feet

   Total Volume = $Slab_{vol}$ + $2F_{1\,vol}$ + $2F_{2\,vol}$
   800 ft³ = $Slab_{vol}$ + (2 × 30 ft³) + (2 × 20 ft³)
   800 ft³ = $Slab_{vol}$ + 100 ft³
   800 ft³ − 100 ft³ = $Slab_{vol}$
   700 ft³ = $Slab_{vol}$
   For cost of total volume, convert ft³ to yd³ = 800 ft³ ÷ 27 ft³/yd³ = 30 yd³ (rounded up)
   Cost = 30 yd³ × $87 = $2,610

"Everyone sit down. I'm going to hand out copies of the foundation plans. I don't want anyone to look at this and freak out on me, telling me they can't figure anything out. We'll go over it together." Mr. Whyte passed papers out to the class.

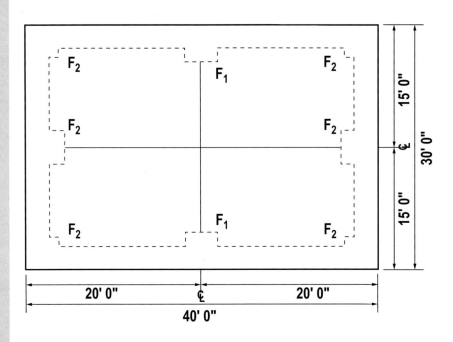

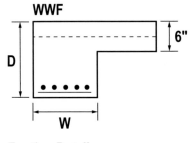

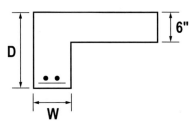

Footing Details

Slab Edge Detail

W = 8"   D = 16"

"Look at the outline of the slab," Mr. Whyte continued. "At the bottom it says 40 feet 0 inches. That's the length. What's the little CL above the 40 feet mean?"

"Um. That's the centerline, Mr. Whyte," Tom said.

"Exactly. It marks the middle of the building, with 20 feet on either side. Look on the right edge of the slab. What's that say, Jorge?"

"30 feet 0 inches. It has the centerline marked, too."

"Good. So we know that the slab is 40 feet long and 30 feet wide. Everyone with me so far?"

The class nodded.

"Look at the broken line around the slab. That marks the footings. There will be footings running around the entire edge of the slab." Mr. Whyte went on.

"Why are there so many F_2s, Mr. Whyte?" Al asked.

"That's next, Al. Everyone look on the drawing and find the two places where it says F_1. Found 'em? Good. Now find the six places that say F_2."

The students studied their diagrams, counting as they found each F_2. Jorge turned to Olivia and shook his head. Olivia pointed to the diagram and Jorge thanked her.

"Everyone found all of them? Good. Now we need to study the footing detail diagram in the lower left corner of the plans. The footer is shaped like this and the slab is 6 inches thick."

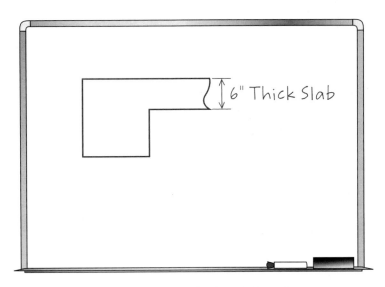

"See the dotted line through the slab? That represents welded wire fabric and it's abbreviated WWF. Welded wire fabric is used to reinforce the slab and make it really strong." Mr. Whyte added a dotted line to his drawing on the board.

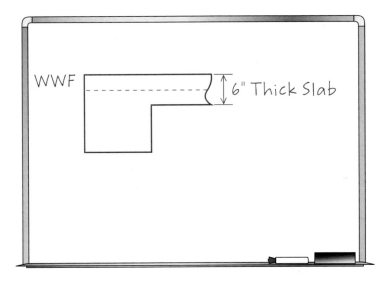

"Look at the bottom of the footing drawing. These dots represent rebar—rebar is reinforcing rod that's also used to make the slab stronger. Rebar is used in the footings."

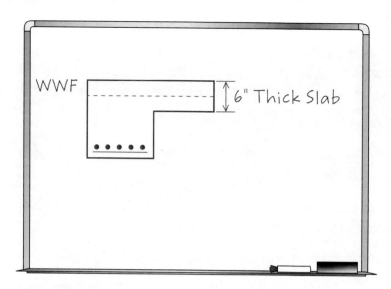

"This is going to be one strong slab, Mr. Whyte," Al said.

"It is, Al. It is. This slab is really overkill for a home shop, but Mr. Brown wants a slab that's strong enough to support a vehicle. Back to volume. We don't need to know about welded wire fabric and rebar to calculate volume, so you can ignore all of that information. What we do need to know is the length, width, and depth of the footings. Look on the foundation plan. There's a chart below the words Footing Details. See where it says F_1 4 feet 0 inches by 4 feet 0 inches by 2 feet?"

"Yeah," Phil said. "Does that mean every time we see F_1 we dig a hole 4 feet long, 4 feet wide, and 2 feet deep?"

"You bet, Phil," Mr. Whyte answered.

"And I bet that every time we see F_2, it means to dig a hole 2 feet long, 2 feet wide, and 1 foot 4 inches deep," Al said.

"You bet right, Al," Mr. Whyte said. "These aren't just any holes. These are nice uniform holes with 90-degree edges. So what geometric shape would that make the hole? Travis, any ideas?"

Math Speak

A cube is a three-dimensional object in which all sides form squares. A rectangular solid is a three-dimensional object in which all sides form rectangles.

"Well," Travis began slowly. "The length and width are equal. F_1 is 4 feet square and F_2 is 2 feet square, but the depths don't equal the width and length, so I know it's not a cube, 'cause all of the dimensions are equal in a cube."

"So far, so good. Anything else?"

Travis thought for a moment, then smiled. "It's called a rectangular solid."

"Excellent. And how, may I ask, do you calculate the volume of a rectangular solid? Sandy?"

"Length times width times depth," Sandy answered.

"Length times width times depth," Mr. Whyte repeated as he drew the footing diagram on the board. "F_1 is 4 feet by 4 feet by 2 feet, and F_2 is 2 feet by 2 feet by 1 foot 4 inches."

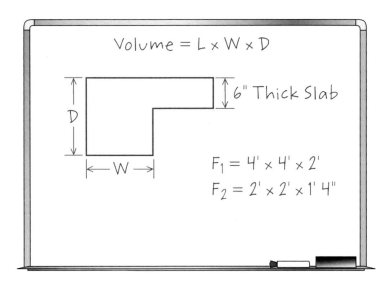

"Mr. Whyte," Al said. "This is getting confusing."

"It's getting confusing because you're trying to look at everything at once, Al. Look at it this way." Mr Whyte scribbled on the board. "We have a slab. It's 30 feet by 40 feet and 6 inches thick—that's ½ a foot—so the slab volume is 600 cubic feet. Now we have two footings—F_1—that are 4 feet by 4 feet by 2 feet. See?"

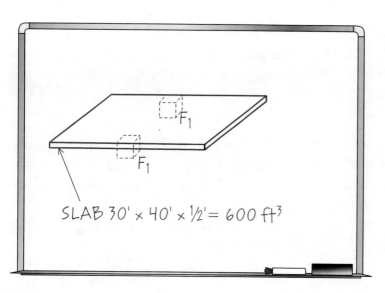

"Uh, yeah, except...," Al began.

"Except what, Al?" Mr. Whyte asked.

"Mr. Whyte, that slab is sitting right on top of the footing. We already calculated the slab, so if we use 2 feet as F_1's depth, it'll be too much."

Mr. Whyte smiled as he wrote on the board. "You got it, Al. For us to figure out the correct volume of F_1, we need to subtract the thickness of the slab, which is 6 inches."

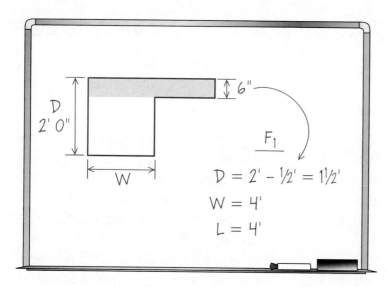

Mr. Whyte turned back to the class. "Since 6 inches equals ½ foot, the depth of the footing is 1½ feet. With a depth of 1½ feet and a length and width of 4 feet, what's the volume? Olivia? Come to the board and complete the problem."

Olivia hurried to the front of the class and wrote on the board. "Volume equals length times width times depth. In this case, it's 4 times 4 times 1½. Four times 4 is 16. Now I'm going to convert the 1½ to a fraction, which is 3/2. Multiply 16 by 3/2, and you get 48/2. Reduce 48/2 to 24, so the volume of F_1 is 24 cubic feet."

$$F_1 \text{ Volume} = L \times W \times D$$
$$= 4' \times 4' \times 1\tfrac{1}{2}'$$
$$= 16 \times 1\tfrac{1}{2}$$
$$= 16 \times \tfrac{3}{2}$$
$$= \tfrac{48}{2}$$
$$= 24 \text{ ft}^3$$

"Good job," Mr. Whyte said. "Everyone write that down—F_1 volume equals 24 cubic feet. Al, think you can figure out the volume for F_2 for us?"

Al went to the board. "F_2 volume is length times width times depth. I have the length and width. They're both 2 feet. The depth on the drawing is 1 foot 4 inches, but that includes the slab, so I have to subtract the slab thickness of ½ foot to get the correct depth. Shoot, I have to convert everything to inches. One foot 4 inches equals 16 inches. The slab is ½ foot thick, which is 6 inches. So the depth of F_2 is 10 inches. F_2's dimensions are 2 feet times 2 feet times 10 inches."

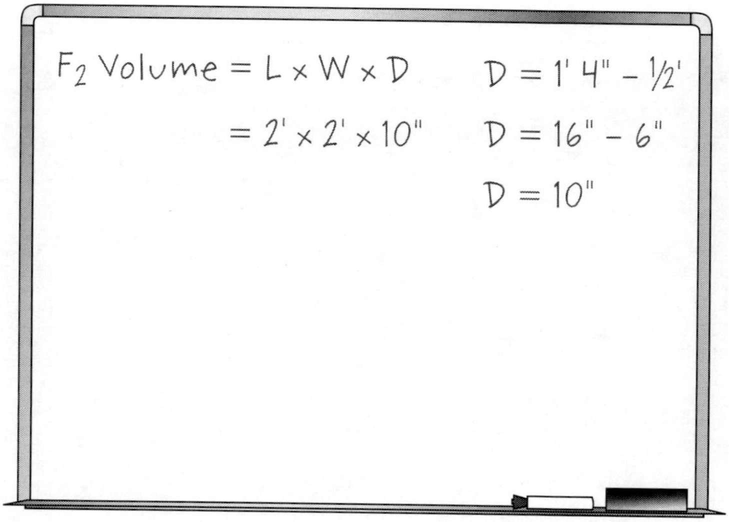

F_2 Volume = L × W × D   D = 1' 4" – ½'
         = 2' × 2' × 10"   D = 16" – 6"
                                    D = 10"

Al went on. "Now I have to convert again. Okay. 10 inches is $^{10}/_{12}$ feet. There's a nice round number!"

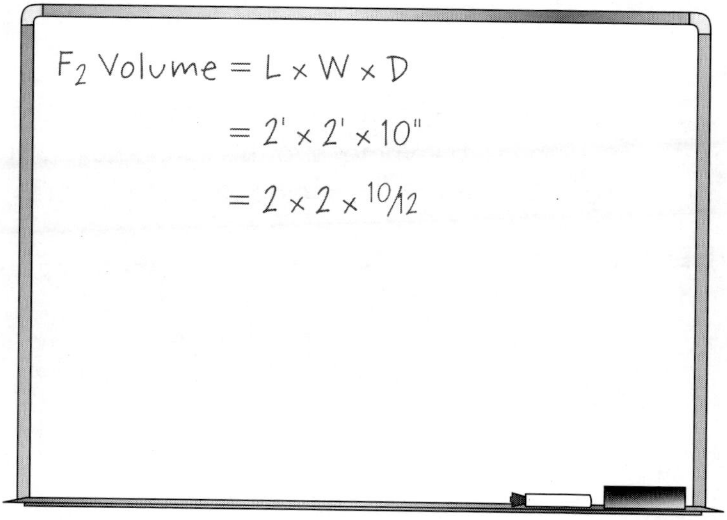

F_2 Volume = L × W × D
         = 2' × 2' × 10"
         = 2 × 2 × $^{10}/_{12}$

The class laughed and Al pressed on.

"Now I multiply 2 times 2 times $^{10}/_{12}$. That equals 4 times $^{10}/_{12}$, which is equal to $^{40}/_{12}$. Hot dog, I get to reduce, too! When I divide 40 by 12, I get 3 with 4 left over. That's $^{4}/_{12}$, and I can reduce the $^{4}/_{12}$ to ⅓, so I get 3⅓ cubic feet for the volume of F_2," Al said.

$$F_2 \text{ Volume} = L \times W \times D$$
$$= 2' \times 2' \times 10''$$
$$= 2 \times 2 \times {}^{10}/_{12}$$
$$= 4 \times {}^{10}/_{12}$$
$$= {}^{40}/_{12} \longrightarrow 12\overline{)40} \;\; {}^{3\;4/_{12}}$$
$$\phantom{= {}^{40}/_{12} \longrightarrow 12\overline{)}}\underline{36}$$
$$\phantom{= {}^{40}/_{12} \longrightarrow 12\overline{)40}}\;4$$
$$= 3\,{}^{4}/_{12} = 3\,{}^{1}/_{3} \text{ ft}^3$$

Mr. Whyte laughed. "Excellent, Al. Take a bow."

Al snapped to attention and bowed to the class. "Thank you, thank you."

"Okay, Al. You can take a seat," Mr. Whyte said as he turned to the whiteboard and drew a diagram of the slab. "We have the slab volume, and the volume of both of the footings. What other volume do we need? Tom?"

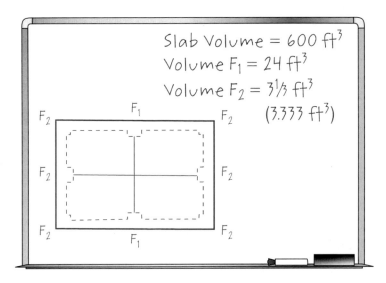

"We need the volume of the footings on the edge of the slab," Tom said.

"All right. We have a little wrinkle here. We have F_1 and F_2 footings in the middle of the slab edge footings. What are we going to do about them?"

"Well," Olivia began slowly. "We already used that part of the foundation so we can't include it in the slab edge calculation. We need to subtract, but I'm not sure what to subtract."

"On the bottom edge we need to subtract 2 feet for the length of the F_1 footing on the left, and 4 feet for the F_1 footing in the middle, and then another 2 feet for the F_2 footing on the right," Sandy said.

"Good, Sandy," Mr. Whyte said. "Did everyone get that?" Mr. Whyte drew on the board. "The lower edge of the diagram is the length. It's 40 feet, and we have two F_2 footings and one F_1 footing in it, so we subtract the length of each of the footings on that edge of the slab from the total length of the slab. And we get a length of 32 feet."

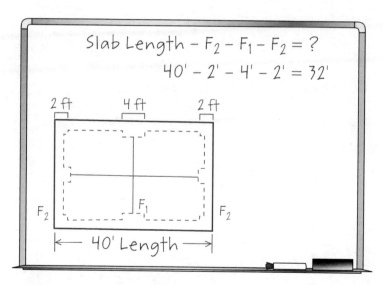

"Mr. Whyte," Jorge asked. "Could we have multiplied the length of F_2 by 2 instead of subtracting it twice?"

"Sure could," Mr. Whyte said. "What Jorge is saying is to write the equation as slab length minus 2 times F_2's length minus F_1's length, and you still get 32."

Mr. Whyte drew another diagram. "What's the volume of the footing on the bottom edge of the diagram?" Mr. Whyte asked. "I want you to set up the problem completely. Start at the beginning with a formula—use variables and show me all of your work. You'll need to study the slab edge detail diagram. Who feels confident?"

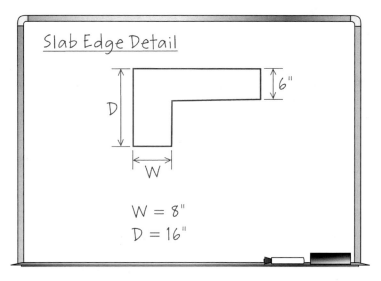

"Hey, I'm hot today, Mr. Whyte. I'll give it a go," Al said, returning to the board. "To calculate volume, you need to know the length, width, and depth."

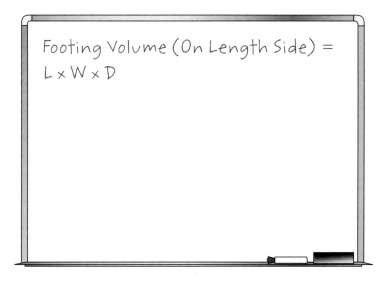

"The tricky part here is what length, width, and depth. I need to look at the plans," Al said as he studied the plans. "There are three footings in the length—two F_2s and one F_1—I need to subtract the length of those from the total slab length just like Sandy said. The length is the slab length minus 2 times the length of F_2 minus the length of F_1 or 32 feet." Al wrote on the board and then looked at Mr. Whyte. "So far, so good?"

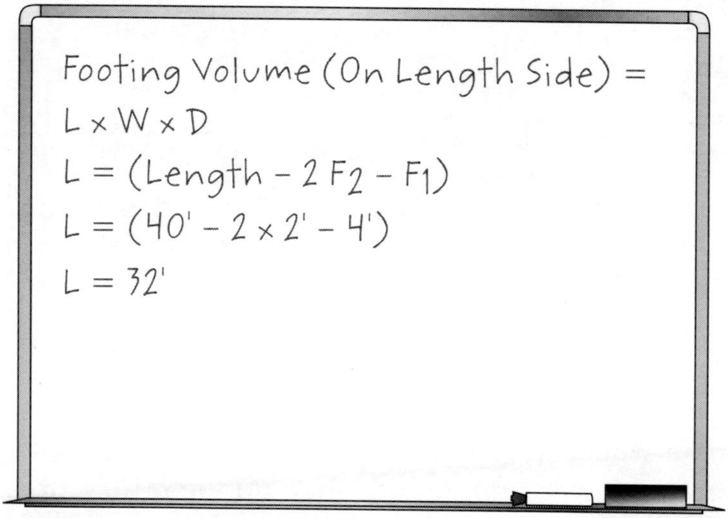

$$\text{Footing Volume (On Length Side)} = L \times W \times D$$
$$L = (\text{Length} - 2\,F_2 - F_1)$$
$$L = (40' - 2 \times 2' - 4')$$
$$L = 32'$$

"You're cookin' now, Al."

"The footing width is easy. It's 8 inches—I got that from the slab edge detail diagram." Al continued. "Now for the depth. The slab edge detail diagram shows the depth as 16 inches, but that includes the slab depth, and we already used that, so I need to subtract it from the footing depth."

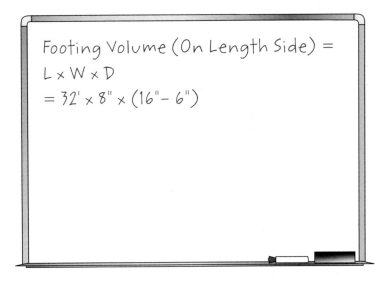

"Good show, Al," Mr. Whyte said. "Can someone tell me why Al put the equations for the length and depth in parentheses?"

"'Cause he wanted us to do those first," Jorge said.

"That's true, Jorge. If he hadn't put them in parentheses, he would have had to multiply F_1 times the slab edge width by its depth, and then do the subtraction. That would be wrong but why?"

The students studied the writing on the board. Some shook their heads, while others tapped their pencils. Olivia whispered to Tom. Tom tilted his head, thinking. Then he nodded to Olivia.

"I think Olivia's got it, Mr. Whyte."

"What do you think, Olivia?"

"Well...the formula for volume is length times width times depth. To get the length of that footing, you need to subtract the lengths of the three footers because you already used them in the footer equations."

"Very good, Olivia," Mr. Whyte said. "Okay, Al. Finish the problem."

"First, I subtract the 6 inches from the 16 inches, and then I convert inches to feet and voila! It's 17.7 cubic feet."

Footing Volume (On Length Side) =
L × W × D
= 32' × 8" × (16" − 6")
= 32' × 8" × 10" = 32' × $^8/_{12}$' × $^{10}/_{12}$'
= 32' × $^2/_3$' × $^5/_6$'
= 32' × $^{10}/_{18}$' = $^{320}/_{18}$
= 17.7 ft^3 (or 17$^{14}/_{18}$ = 17$^7/_9$)

$$18\overline{)320.0} = 17.7...$$

"Well done, Al," Mr. Whyte said. "So far, we have the slab volume, the volumes of F_1 and F_2, and the volume of the footings on the slab length. Do we need anything else?"

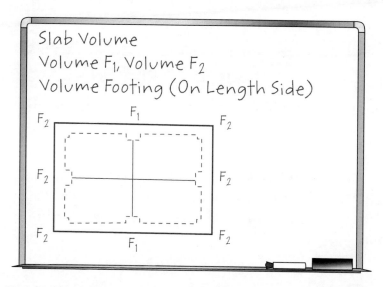

Slab Volume
Volume F_1, Volume F_2
Volume Footing (On Length Side)

"Yeah, the volume of the footings on the width of the slab," Phil said.

"You got it. I'm going to let you guys finish this problem for homework. But before I let you go, let's talk about how many F_2 lengths you need to subtract from the slab width."

"Looks like one, Mr. Whyte," Phil said. "We already used the corner footings to figure out the volume of the footings on the length of the slab."

"Nope. Think," Mr. Whyte said, drawing on the board. "The length and width dimensions for F_2 are both 2 feet, right?"

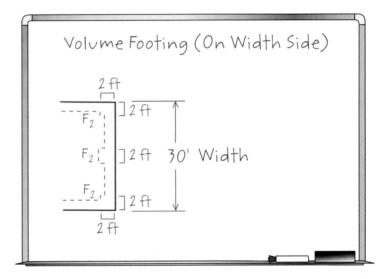

"Oh, yeah," Phil said. "On the corner footings, there's 2 feet on the length side of the slab and 2 feet on the width side of the slab, so we need to subtract all three footings."

Just then the bell rang. "Finish the problem for homework," Mr. Whyte called out to the students as they filed out of the room. "And make sure you show all of the work."

Chapter 7 Review Problems

1. Take a minute and complete the problem Mr. Whyte assigned for homework. You need to calculate the footing volume for the width of the slab, and then calculate the total volume. After you have calculated the total volume, round up to the nearest whole number. (Hint: Study the slab edge detail diagram. Remember to subtract the depth of the slab from the depth of the footing before you calculate the volume of the footings.)

   Total volume = 600 ft³ + (2 × 17.7 ft³) + (2 × 13.3 ft³) + (2 × 24 ft³) + (6 × 3.3 ft³) = 730 ft³ (rounded up)

2. A foundation has a slab volume of 650 cubic feet and a footing volume of 90 cubic feet. Use the variables S_{vol} for the slab volume and F_{vol} for the footing volume.

   a. Write the equation for calculating the total foundation volume.

   Total volume = S_{vol} + F_{vol}

   b. Substitute the values given in the equation.

   Total volume = 650 ft³ + 90 ft³

   c. Calculate the total volume of the foundation.

   Total volume = 740 ft³

3. A foundation slab is 20 feet by 30 feet, and is 6 inches thick. The footings are uniform around the edge of the slab and are 8 inches wide and 12 inches deep.

   a. Define variables for the slab and footing volumes.

   Slab volume = S_{vol} Footing volume = F_{vol}

   b. Using the variables defined in step a above, write the equation for calculating the total foundation volume.

   Total volume = S_{vol} + F_{vol}

   c. Calculate the slab volume.

   S_{vol} = 20' × 30' × 6/12' = 300 ft³

   d. Calculate the footing volume.

   F_{vol} = [2 × 30' × 8/12' × 6/12'] + [2 × (20 – 16/12') × 8/12' × 6/12'] = 33 ft³ (rounded up to whole yard)

e. Calculate the total foundation volume.

   Total volume = 300 ft³ + 33 ft³ = 333 ft³ (rounded up)

4. The total foundation volume is 830 cubic feet. The total volume of the footings is 130 cubic feet.

   a. Write an equation to solve for the total foundation volume. Here's a hint:

   Total foundation volume = $S_{vol} + F_{vol}$

   b. Fill in the known variables. (Hint: You don't know the volume of the slab.)

   830 ft³ = S_{vol} + 130 ft³

   c. Solve for the unknown variable.

   830 ft³ − 130 ft³ = S_{vol} + 130 ft³ − 130 ft³ (subtract 130 from both sides)

   700 ft³ = S_{vol}

5. A slab is 20 feet by 20 feet and is 6 inches thick. There are no footings. Solve for slab volume.

   Slab volume = 20' × 20' × 6/12'

   Slab volume = 400 sq ft × ½'

   Slab volume = 200 ft³

6. Write a short step-by-step procedure describing how to find the total volume of an irregularly shaped object. (Hint: A procedure is short, simple, and direct. It tells you how to do something, but it doesn't tell you why you are doing it unless you need to know the information to complete the procedure. Start with the sentence: To calculate the complete volume of an irregularly shaped object, perform the following procedure...)

   Answers will vary, but should follow the basic procedure provided in this chapter.

Careers in Construction—Structural Engineer

Mr. Brown had a structural engineer design the foundation for his shop. Structural engineers specialize in determining how much stress something can withstand before being damaged. Mr. Brown told the engineer that the shop might be used for vehicle maintenance, so the engineer designed a slab with reinforcing fabric in it.

To be a structural engineer, you must go to college and excel at math. Structural engineering is applied to all sorts of things. Did you ever sit in a chair that collapsed under your weight? That's a structural engineering failure. The Gateway Arch in St. Louis is a marvel of structural engineering. A tram runs from the base of the arch all the way to the top. Millions of people ride through the legs of the arch to and from the top every year. Not a lot of room for failure there!

 Terms

Frostline: The depth at which the soil freezes in a certain area.

Sill plate: The lowest horizontal board of a wall. Its purpose is to secure walls to the floor. These boards are usually 2 × 4s or 2 × 6s. In exterior walls, the sill plate is attached to the foundation with anchor bolts. In interior walls, the sill plate is attached to the floor. The nail studs are then nailed to the sill plate.

Slab-on-grade: A monolithic slab. This means that the concrete for the slab and footings is poured at the same time.

 Did You Know

Frost-Protected Shallow Foundation (FPSF)

In cold areas, footings are typically located below the frostline for the area. This is because as water freezes, it expands. When water is in the soil and it freezes, it pushes against the footings and can actually move them! This can cause the building foundation to crack. To prevent this, footings are located below the frostline. (If the ground doesn't freeze, the water in the soil doesn't expand, so the footing doesn't move.)

With a frost-protected shallow foundation (FPSF), insulation is placed around the foundation. This prevents heat loss from the foundation and actually makes the soil around the foundation warmer, thus preventing freezing. Using an FPSF saves on excavation costs, as well as material costs (less cement is needed), but be aware! This is new technology and may need a special review at the local codes enforcement office.

Chapter 8
Choosing Teams

In this chapter, students must make linear measurements quickly and accurately. Mr. Whyte used lumber to demonstrate the concept of fractional measure, but you can use paper that is marked ½, ¼, ⅛, etc. Use the same method that Mr. Whyte used in the text.

Students must also calculate perimeters. Students use the building perimeter to estimate the number of sill plates and studs needed in the exterior wall of a building. There are tricks to all trades. These tricks are used by tradespersons so often that they often no longer know why they use them—just like Mr. Conrad. One trick that construction estimators often use is to allow one 2 x 4 stud per foot of perimeter. Since each corner needs at least three studs and each opening needs two on each side, using the one-per-foot trick compensates for these additional studs without going into a lot of math.

Point out to the students that to add or subtract fractional values, they must convert them to common denominators, so ½ + ¼ = 2/4 + ¼ = ¾.

NOTES

Chapter 8

Choosing Teams

"Okay, guys. Let's step into the shop to talk about the day's work," Mr. Whyte said to the group of students standing outside Mr. Brown's shop.

Mr. Whyte headed toward Mr. Brown's shop. The class completed work on the shop's slab three weeks ago, and then a crew came down from Georgia and built the whole shop in three days! It was a metal building, and it went up just like a giant erector set. Mr. Brown let the class use the building, so they had a secure place to store their tools.

"Hey, Mr. Whyte," Al called as the group made their way to the shop. "Where's Sandy?"

"Don't know, Al. Her mother didn't call the office to let us know about her being absent."

Phil turned to Olivia and said, "Do you know where she is? She told me she was really looking forward to working today." The class finally had the house foundation done, and they were ready to start framing the exterior walls. Everyone was excited.

Olivia shook her head. "Maybe she's just late. Maybe her mom will drop her off here."

"Listen up," Mr. Whyte said when everyone got into the shop. His voice echoed in the big empty building. "Before Mr. Conrad gets here, I want to make sure that all of you can accurately read measurements."

Mr. Conrad was a foreman for JM Custom Homes. Since Mr. Whyte couldn't be on the construction site full time, the owner of the construction company, Jackson Moran, said Mr. Conrad could help out with the house construction. Mr. Moran liked Mr. Whyte's home construction program a lot. He often hired graduates of the program to work for his company. Mr. Conrad was also Tom's dad. Tom was really worried about working for his father.

Mr. Whyte moved to the center of the floor and stood next to a pile of 2 × 4s that had been cut to different lengths. "Lots of people have trouble reading fractional inches from a ruler, but construction workers need to be really good at it. And what's the only way to get good at something?" Mr. Whyte asked.

"PRACTICE!" the class shouted.

The sound echoed off the walls of the empty building, startling Mr. Whyte, who ducked. "Gosh, you guys could knock me down in here! Right. Practice. Everyone pull out your measuring tape."

The sound of a car door slamming penetrated the building. Everyone looked expectantly at the shop door. When it opened, Sandy walked in.

"Sorry I'm late, Mr. Whyte," Sandy murmured.

"Glad you could make it, Sandy. Get your tape measure out and join us."

Sandy hurried over to the only furniture in the shop—a battered old workbench—picked up her tool belt, and joined the rest of the students sitting on the floor.

"A tape measure is a tool," Mr. Whyte said. "And just like any other tool, you need to know how to use it. The first purpose of a tape measure is to accurately measure linear distances. Linear means straight. The tape measure manufacturers know a little something about how tape measures are used." Mr. Whyte held up his tape measure and pulled a couple of feet of tape out of the case. "The front side of the tape is marked in inches and feet. On my tape, the top line of the tape is marked in 16ths of an inch, and the bottom is marked in 32nds of an inch. On the top line, each foot is marked with a big arrow, so the top line is marked feet and inches and the bottom is marked in just inches. See how my tape has 16 inches marked with a red box around it? That's because 16 inches is a common measurement in construction."

The students studied the markings on the front of their tapes. Al looked over at Phil's tape and said, "I feel kinda cheated. My tape doesn't have 32nds marked on it. What's that stuff written on the back of your tape, Mr. Whyte?"

"I have a deluxe model, Al. The back of my tape has a handy little table of fraction-to-decimal equivalents, nail sizes, some formulas, and a bunch of other stuff. Everyone look on both sides of your tape to see what information is on it."

Everyone looked at their tapes and then looked at their classmates' tapes. Jorge pointed to Al's tape, laughed, and said, "Guess you got the no-frills tape, Al."

"Man, if anyone deserves one of those fraction-to-decimal converter things, it's me!" Al said with a shake of his head.

Mr. Whyte pulled a long board from the pile of lumber next to him. "Look at this board. This is a whole."

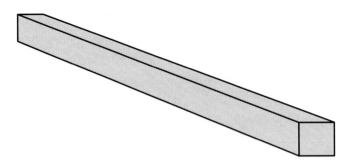

"A whole what, Mr. Whyte?" Phil asked.

"A whole anything. A whole inch, whole foot, whole yard, or whole mile. It represents a whole of any linear measure," Mr. Whyte answered as he pulled a shorter piece of wood from the pile. "This is a half." Mr. Whyte clapped the shorter board on top of the longer board. "It's ½ the length of the whole."

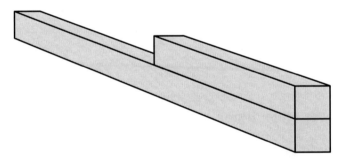

Mr. Whyte continued. "Because it's ½ the length of the long board, I can put it anywhere on the long board, and I'll still have ½ left over."

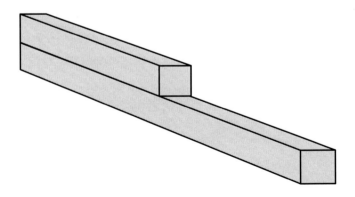

Mr. Whyte pulled another piece of wood from the pile. This one was shorter than either of the other boards. "This is a ¼." Mr. Whyte slapped the shorter board on top of the other two boards. "See, it's ¼ the length of the whole board. And it's ½ the length of the half board."

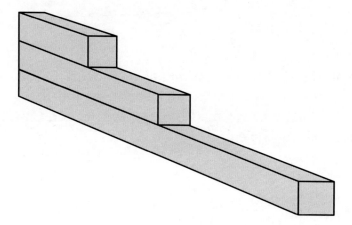

Mr. Whyte pulled yet another piece of wood from the pile. "This is another ¼." Mr. Whyte shoved the board end-to-end with the shortest board. "See, two ¼s make a half."

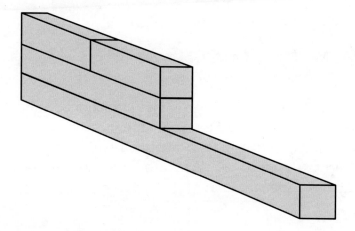

Then he placed the three shorter pieces of wood end-to-end on the longest piece. "See, two ¼s and one ½ make a whole. Do you guys see where I'm going with this?"

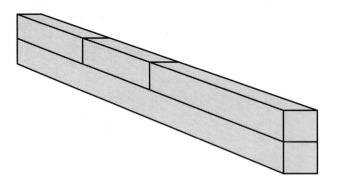

"I think so, Mr. Whyte," Al said as he moved the three short pieces of wood around. "You're saying that no matter which way we mix 'em up, two ¼s and one ½ always make a whole."

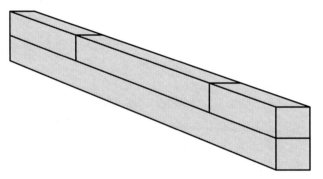

Mr. Whyte nodded.

"Okay, Mr. Whyte, I got it," Tom said. "But what does this have to do with measuring?"

"What I'm trying to illustrate is that a fractional amount can occur anywhere. Look at your tape measure. If I said, 'Find ½ inch,' you would probably have no problem locating the ½ inch." Mr. Whyte pointed to the ½-inch mark on his tape measure.

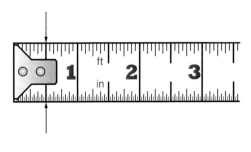

– 8.6 –

"Or ¼ inch." Mr. Whyte moved his thumb closer to the end of the tape.

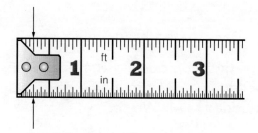

"Or even ¾ inch." Mr. Whyte moved his thumb again.

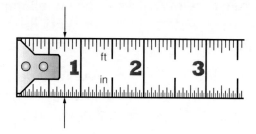

"But what if I said, 'Drill a hole in this board every 1½ inches?'" Mr. Whyte asked. "Could you do it?"

Tom studied his tape measure. "Sure, every 1½ inches. That's here, here, and here." Tom pointed to three places on his tape measure.

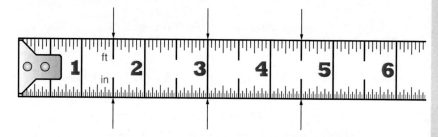

"Good. How did you arrive at those numbers?" Mr. Whyte asked.

"I added 1½ to each number. The first is 1½ inches. 1½ plus 1½ is 3 inches. 3 inches plus 1½ is 4½ inches. If I went on, the next measurement would be 6 inches. 4½ inches plus 1½ inches equals 6," Tom answered.

"Good job. What if I asked you to drill three holes in a board every 1½ inches starting 1¾ inches from the edge? Could you mark the board correctly?"

"Yikes, Mr. Whyte, have a heart," Al said.

The door to the shop opened, startling everyone. Mr. Conrad looked in at the group and gave them a quick smile. He nodded to Mr. Whyte.

"Here's Mr. Conrad," Mr. Whyte said as he turned to face the students. "You guys can get to work. Everyone knows the rules: safety glasses on at all times—no exceptions—work with your buddy, and drink lots of water—it's hot out there. Mr. Conrad's in charge."

Mr. Whyte turned to Mr. Conrad. "Our method is simple. We're one team of workers. I've placed my best students in charge of each task. When they're working on that task, they're in charge. When they work on someone else's crew, they're a worker. Phil's crew chief for carpentry, Al's got heating and air conditioning, and Tom's in charge of electrical." Mr. Conrad scribbled in a notebook he was carrying. When Mr. Whyte mentioned Tom's name, Mr. Conrad glanced at his son with a raised brow. Tom's cheeks reddened, and he looked at the floor.

Mr. Conrad looked back at Mr. Whyte and said, "Who's got plumbing?"

Mr. Whyte hesitated. He had intended to give plumbing to Sandy, but now he wasn't sure, so he said, "I haven't assigned anyone yet. We subcontracted the slab plumbing, so I didn't need a crew chief." Mr. Whyte turned to the students. "Okay, guys. Get to work. I want to see some framing when I get back!"

Practice Problems 8-1

Take a moment and do as Mr. Whyte asked. Follow the steps below to complete the problem.

1. Find 1¾ inches on the diagram below, and mark it.

2. Add 1½ inches to 1¾ inches.

   1½" + 1¾" = 2 5/4" = 3¼"

3. Mark the answer for Question 2 on the diagram below.

4. Add 1½ inches to the answer to Question 2.

   3¼" + 1½" = 3¼" + 1 2/4" = 4¾"

5. Mark the answer for Question 4 on the diagram below.

6. Add 1½ inches to the answer to Question 4.

   4¾" + 1½" = 4¾" + 1 2/4" = 5 5/4" = 6¼"

7. Mark the answer for Question 6 on the diagram below.

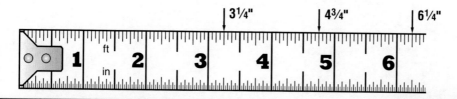

Mr. Conrad led the way outside. The class followed, chattering as they went.

Mr. Whyte stopped Sandy. "I need to talk to you." As Mr. Whyte spoke to Sandy, an old red pickup truck drove onto the property and up to the shop. The truck had a For Sale sign in the back window. "That's Mr. Brown," Mr. Whyte told Sandy. "I need to talk to him. I'll catch up with you later."

Phil nudged Al. "Look at the sign on the truck. It's for sale. That's what you need."

Al stared at the truck. "Maybe," he said.

"Okay, gang," Mr. Conrad shouted. "We're framing today. Mr. Whyte says you guys are pretty good at math, so someone tell me the total length of sill plates I need for a house."

"Oh, Mr. Conrad, that's too easy," Al said. "It's the perimeter of the house."

 **Math Speak**

The perimeter is the distance around a square, rectangle, and triangle. It's usually abbreviated with the letter P. You can calculate the perimeter of these shapes by measuring each side and then adding all of the sides together.

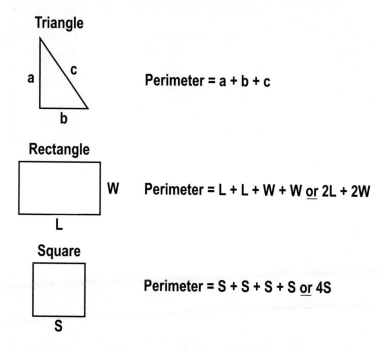

The distance around a circle is called the circumference. It's hard to measure the distance around a circle, but you can easily calculate it using one of the formulas shown below.

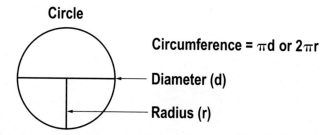

The symbol π is the abbreviation for pi. Pi is pronounced 'pie' just like in apple pie. Pi is the ratio of the circle circumference to its diameter. Pi is always 3.1415926... The numbers after the decimal really continue on forever, but most people round to 3.14 or use the fraction $^{22}/_{7}$.

Practice Problems 8-2

Complete the following problems.

1. Calculate the perimeter of the building shown in the following figure.

   Perimeter = 24' + 24' + 14' + 14' = 76'

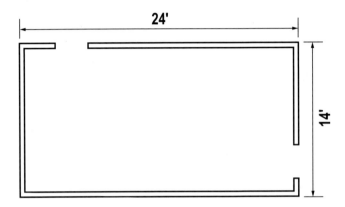

2. Calculate the perimeter of the building shown in the following figure.

   Perimeter = 30' + 30' + 24' + 24' = 108'

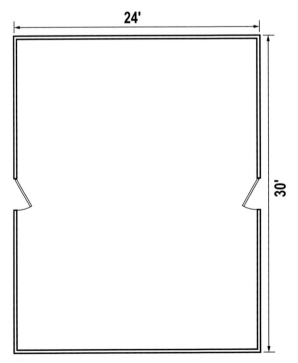

3. Calculate the perimeter of the building shown in the following figure.

Perimeter = 40' + 40' + 30' + 30' = 140'

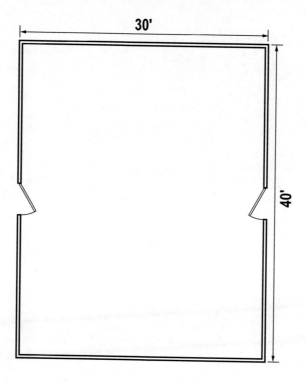

4. In the figure shown below, calculate the length of the wall labeled with the letter a.

   Side a = 30' − 12' = 18'

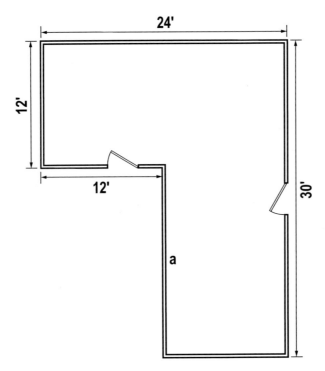

5. Calculate the perimeter of the building shown in the figure following Question 4.

   Last unknown side: 24' − 12' = 12'

   Perimeter = 30' + 24' + 12' + 12' + 18' + 12' = 108'

6. Calculate the lengths of the walls labeled with the letters a and b in the following figure.

Side a = 60' – 20' – 25' = 15'

Side b = 40' – 12' – 20' = 8'

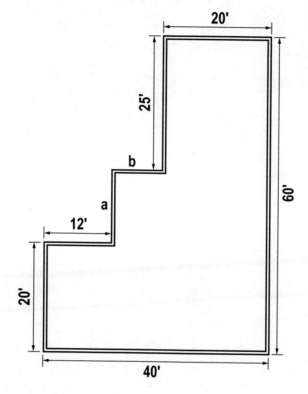

7. Calculate the perimeter of the building shown in the figure following Question 6.

Perimeter = 60' + 20' + 25' + 8' + 15' + 12' + 20' + 40' = 200'

Mr. Conrad continued. "Assuming we're buying 12-foot 2 × 6s for the sill plates, how would you determine the number you need?"

"Me?" Al asked. "I'd look at the **takeoff list**."

Mr. Conrad frowned. "Nice try, but Mr. Whyte told me to make you guys do the calculations."

"You need to divide the perimeter by the length of the sill plates," Tom said quietly.

Mr. Conrad nodded and drew in his notebook. "On a 36-foot by 24-foot building, the perimeter is 120 feet, so I need ten 12-foot 2 × 6s, and then I'll trim them to fit as needed."

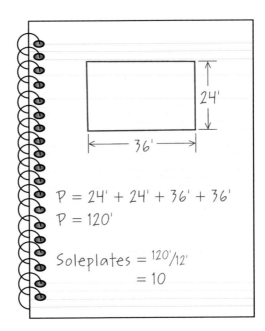

Mr. Conrad continued. "Here's the big question. How do you calculate how many studs you need to frame the exterior wall of a house?"

"Well," Phil began. "First you need to know how far apart you're spacing them."

"16 inches on center," Mr. Conrad said. Mr. Conrad drew a diagram in his notebook. "On center means the center of the 2 × 4. It's abbreviated with the letters OC. And the center of each 2 × 4 is spaced 16 inches apart."

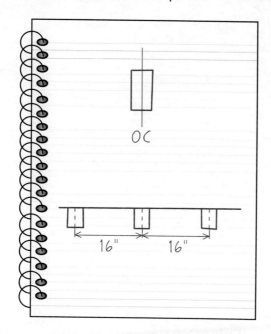

"Tom, how many studs do I need for a 20-foot by 12-foot building, 16 inches on center?"

Tom's eyes opened wide. Then he scribbled in his notebook. "Um, 20 by 12. That's 64. Now multiply times 12 to convert to inches, then divide by 16." Tom looked up. "That's 48 without any adjustments for doors, windows, or corners, or other stuff."

"Show me your work," Mr. Conrad said.

Tom handed his notebook to his dad. Mr. Conrad frowned as he studied it.

Mind Games

When is a 2 × 6 not a 2 × 6? Always!

A 2 × 6 is really 1½ inches by 5½ inches. It starts out at 2 inches by 6 inches when it's cut at the saw mill, but then it's dried in an oven called a kiln. As the lumber dries, it shrinks. Then the lumber is planed to make the sides smooth. By the time the 2 × 6 reaches the building site, it's really 1½ × 5½.

The numbers used to describe the lumber are called the **nominal** size. Here are the measurements for some other commonly used lumber:

2 × 4 is really 1½ × 3½

2 × 8 is really 1½ × 7¼

3 × 6 is really 2½ × 5½

4 × 6 is really 3½ × 5½

But don't worry! A 12-foot length is still 12 feet!

Mind Games

Mr. Conrad used a common trick in construction. When 2 × 4 studs are spaced 16 inches OC, you can calculate the basic number of studs by multiplying the perimeter in feet by ¾. When the stud spacing is 24 inches OC, you can divide the perimeter by two. Both of these tricks work, but Mr. Conrad has been using them for so long, he can't remember the math behind them. Read on as Tom explains it to him.

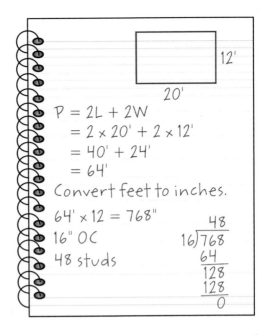

Then Mr. Conrad said, "You came up with the right answer, son. But you did a lot more work than necessary. All you have to do is multiply the perimeter in feet by ¾ to get the basic count. Then, you're right. You need to adjust it for that other stuff."

Al scribbled in his notebook. "Hey, Mr. Conrad, you're right, that does work. But why?"

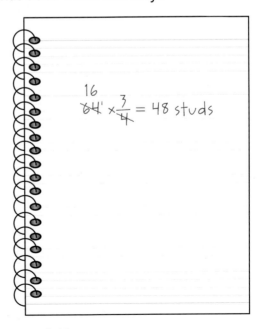

Mr. Conrad looked over Al's shoulder at the figures Al wrote in his notebook. Then he studied the figures in Tom's notebook, scratching his head. After a minute, Mr. Conrad shook his head. "I used to know, Al. But for the life of me, I can't come up with the reason right now. When you're spacing the studs 24 inches on center, all you do is divide the perimeter in feet by two to get the basic count."

The students studied the figures in Tom's notebook and then Al's notebook. Many scribbled in their own notebooks. Some students shook their heads. Then Tom's face brightened. "Of course. 16 inches equals 1 and ⅓ feet or ⁴⁄₃ foot. 24 inches equals 2 feet."

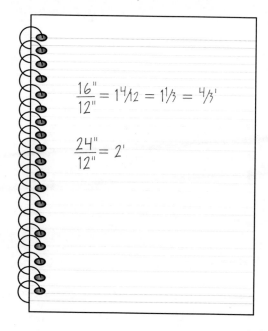

$$\frac{16"}{12"} = 1\ ^4\!/_{12} = 1\frac{1}{3} = \frac{4}{3}{}'$$

$$\frac{24"}{12"} = 2'$$

"What does that have to do with what we're talking about?" Mr. Conrad asked.

Tom's neck reddened, but he pressed on. "By converting the stud spacing to feet, you don't need to convert the perimeter to inches. You can divide the perimeter in feet by ⅔ feet."

Mind Games

Before you perform mathematical functions, be sure to convert all of the numbers to the same units.

Mr. Conrad drew a deep breath and said slowly, "Tom, we are not dividing by ⅓. We are multiplying by ¾. Pay attention."

Tom's neck darkened more. "Dad, look. Dividing by ⅓ feet is the same as multiplying by ¾ foot. When I multiply the perimeter in feet by ¾ foot, I get 48. Then I check my work so I multiply by ⅓ feet, which is the 16-inch spacing in feet, times 48, and I get 64 feet." Tom handed his father the notebook.

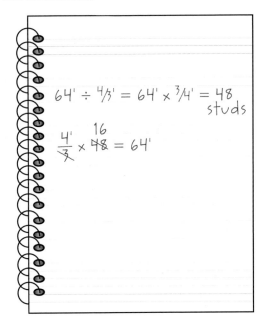

Phil scribbled in his notebook. "Looks right to me, sir."

Al looked at the figures in Phil's notebook and nodded. "Looks good to me, too."

Mr. Conrad stared at Tom, and then he turned to the other students and said, "Okay, everyone. To work." The students moved off toward the foundation site. Only Tom remained. Mr. Conrad looked at the figures in Tom's notebook, lost in thought. He moved his pencil from figure to figure and mumbled, "64 feet…16 inches equal ⅓ feet…"

"Um, Dad," Tom said. "I need my notebook."

"Oh, sure, son. Here you go. Now get to work."

Key To Understanding

When you are working with fractions, remember that dividing by a fraction is the same as multiplying by the **inverse** of the fraction. For example, the inverse of ½ is 2/1, and the inverse of ⅖ is 5/2.

If you divide 1 by ½, the answer is 2 because there are 2 halfs in 1. This is the same as multiplying by 2.

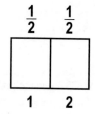

If you divide 2 by ⅖, the answer is 5 because there are 5 ⅖s in 2. This is the same as multiplying 2 by 5/2.

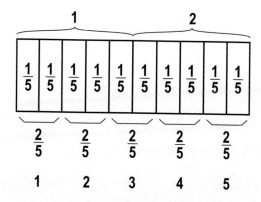

Note that the reverse is true, too. Multiplying by a fraction is the same as dividing by the inverse of the fraction, but most people find multiplication of fractions to be easier than division.

Did You Know

Lumber is sold in board feet. A board foot is a piece of lumber that is 1 inch thick, 12 inches wide, and 12 inches long. When you calculate the volume of a board foot, it comes out to 144 cubic inches.

One board foot volume = L × W × D

= 12" × 12" × 1"

= 144"

A 12-foot 2 × 6 has a volume of 1,728 cubic inches or 12 board feet. (Remember: Always convert all of the numbers to the same units before you begin. There are 144 inches in 12 feet.)

Volume of 12-foot 2 × 6 = L × W × D

= 144 × 2 × 6

= 1,728 cubic inches

To convert 1,728 cubic inches to board feet, divide by 144 cubic inches.

1,728 cubic inches   = 1,728 ÷ 144

= 12 board feet

It takes about 14,000 board feet to build an average size wood-frame house!

Chapter 8 Review Problems

Complete the following problems.

1. On the diagram below, mark 1¼ inches.

2. On the diagram below, mark 3¾ inches.

3. On the diagram below, make three marks on the board every ½ inch starting 1¾ inches from the edge.

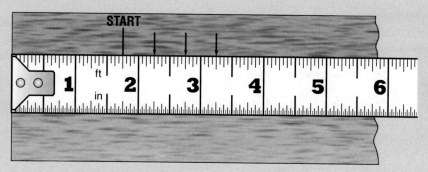

4. On the diagram below, make two marks on the board every ¾ inch starting 2¼ inches from the edge.

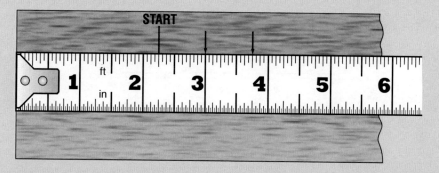

5. On the diagram below, make marks on the board every ¼ inch starting ¾ inch from the edge until you reach 5 inches.

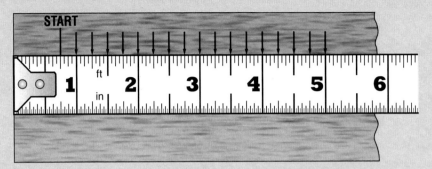

For Questions 6 through 8, first calculate the perimeter of each building, then calculate the number of sill plates needed, and finally calculate the basic number of studs needed.

6. a. P = 24' + 12' + 24' + 12' = 72'

   b. Number of sill plates 12 feet in length = **72'/12' = 6**

   c. Basic number of studs 16 inches OC =

   72' ÷ 16/12 = 72' × 12/16 = 54 (invert 16/12 and change ÷ to ×)

   d. Basic number of studs 24 inches OC =

   72' ÷ 24/12 = 72' ÷ 2' = 36 (reduce 24/12 to 2 and divide or invert 24/12 and change ÷ to ×, and then reduce the 12/24 to ½—the answer is the same)

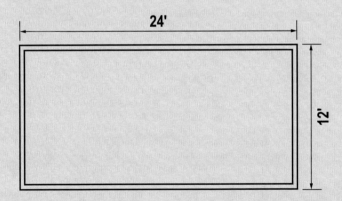

7. a. P = **36' + 24' + 12' + 12' + 36' − 12' + 24' − 12' = 120'**

   b. Number of sill plates 12 feet in length = **120'/12' = 10**

   c. Number of sill plates 10 feet in length = **120'/10' = 12**

   d. Basic number of studs 16 inches OC =

   120' ÷ ¹⁶⁄₁₂' = 120' × ¹²⁄₁₆' = 90 (invert ¹⁶⁄₁₂ and change ÷ to ×)

   e. Basic number of studs 24 inches OC =

   120' ÷ ²⁴⁄₁₂' = 120' ÷ 2' = 60 (reduce ²⁴⁄₁₂ to 2 and divide or invert ²⁴⁄₁₂ and change ÷ to ×, and then reduce the ¹²⁄₂₄ to ½—the answer is the same)

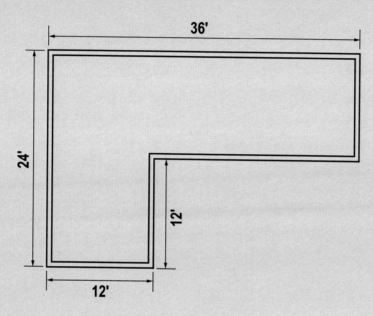

8. a. P = 120' + 60' + 120' + 24' + 24' + 60' = 408'

   b. Number of sill plates 12 feet in length = **408'/12' = 34**

   c. Number of sill plates 10 feet in length = **408'/10' = 40.8 round to 41**

   d. Basic number of studs 16 inches OC =

   $408' \div {}^{16}\!/_{12}' = 408' \times {}^{12}\!/_{16}' = 306$ (invert ${}^{16}\!/_{12}$ and change ÷ to ×)

   e. Basic number of studs 24 inches OC =

   $408' \div {}^{24}\!/_{12}' = 408' \div 2' = 204$ (reduce ${}^{24}\!/_{12}$ to 2 and divide or invert ${}^{24}\!/_{12}$ and change ÷ to ×, and then reduce the ${}^{12}\!/_{24}$ to $\frac{1}{2}$—the answer is the same)

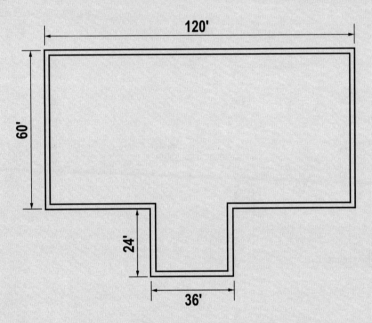

9. Mr. Whyte stressed safety when he sent the students out to work on the house. Safety is important all the time. Think of a time when you or someone you know had an accident or near miss because of an unsafe act, and then write a short description of what happened and how you think the accident could have been prevented.

   Answers will vary. Have each student read their answer aloud and encourage discussion.

History—Forestry

Forestry today is managed. That means trees that are cut for lumber are selected based on their size and age. When trees are cut, new trees are planted to replace them. It wasn't always that way. In the late 1800s, some natural forests were clear-cut. That means that all the trees were cut regardless of size. The cutting was so severe and widespread that some species were completely wiped out! Back then trees were usually transported to sawmills by floating them down a river. Men with long poles pushed and prodded the logs to keep them moving. (Did you ever wonder where the word logjam came from? While the logs were floating down the river, they would sometimes turn and get stuck, then the logs would pile up behind the stuck log—hence, the term logjam!) Today, some lumber companies retrace the rivers that were used to transport logs looking for logs that may have sunk. Scuba divers harvest these logs and then the logs are sawed, dried, and used to build things. Sometimes they even find a log of a species that was wiped out!

 Terms

Inverse: Opposite. In math, the inverse of a function is found by swapping the two variables. For example, the inverse of ½ is ⅔.

Nominal: In name only. The name 2 × 4 implies that the board is 2 inches by 4 inches, but it's not. It's really 1½ inches by 3½ inches, so a 2 × 4 is a 2 × 4 in name only.

Takeoff list: A list of materials needed to complete the project. The list is compiled by analyzing the project drawings.

Careers in Construction—Carpenter

The duties of a carpenter can vary from job to job. Carpenters who work on commercial buildings may work with concrete, steel, and prefabricated building materials. Carpenters who work on residential buildings will most likely work with wood. In the construction industry, there are two broad categories of carpentry: rough and finish. Rough carpentry includes framing—just like the framing that Mr. Whyte's class is doing. Finish carpentry includes building stairs (like the beautiful staircase shown in the picture) and installing doors, cabinets, and wood paneling.

Some carpenters even build furniture like this unique corner hutch that's made from an old window purchased from a salvage yard.

Carpentry requires a broad set of skills that includes math—big surprise, huh? You can break into carpentry by being an apprentice to a skilled carpenter or attending a community college or other training program. Since we always need new homes and commercial buildings, as well as repairs to older homes and buildings, there is always a need for skilled carpenters.

Did You Know

In the United States, we use the English system for linear measurements. This system uses inches, feet, yards, and miles. The English system is sometimes called the Imperial system or the American system or sometimes even the Customary system.

Other countries use the metric system. The metric system makes linear measurements in meters (one yard is equal to 0.90 meter). This system uses powers of ten to express large numbers, small numbers, and fractions. Many people think that the United States should use the metric system, but so far it is not widely used.

NOTES

Chapter 9
Gravity Can Work for You or Against You

This chapter explains how to calculate the correct length of piping between fittings. You may use various lengths of PVC piping and fittings to set up sample problems for the students to solve.

This chapter uses right angle trigonometry to calculate the travel, run, or offset when laying out piping. It also uses trigonometry to calculate grades.

While discussing grades, tell the students that this is an exception to the rule to convert all measurements to like units before multiplying or dividing, because you want the answer to be in inches per foot. This concept can be difficult for students; try relating it to a dozen eggs for $1. Write on the board 1 dozen/$1, and then write 1 dozen/$1 × $5 = 5 dozen. Point out that $5/$1 = 5 (the dollars cancel). Next, write 1 inch/foot, and then write 1 inch/foot × 5 feet = 5 inches. Point out that the feet cancel. Finally compare the two equations, relating the similarities.

NOTES

Chapter 9

Gravity Can Work for You or Against You

"You know I can do it, Mr. Whyte," Sandy said. Sandy and Mr. Whyte were in Mr. Brown's shop standing near the workbench. The door to the shop was open, so the interior was bright with sunlight. They could hear the voices of the other students working outside at the house site.

"That's not in question, Sandy. Of course, you can do it. The question is will you be able to finish the job? Your mind isn't on your work, and you're absent a lot. If I assign you as crew chief for plumbing, you need to be there to answer questions."

Sandy stood up straight, took a deep breath, then said, "My mom's doing a lot better now. I talked to her, and my grandmother talked to her, too. Grandma told Mom I was acting more adult than she was, and it wasn't fair that she expected me to pick up the slack at home while my Dad works in Hong Kong. Mom's doing okay. She doesn't need me to be home with her anymore."

Mr. Whyte twisted the end of his mustache while he studied Sandy. Then he looked at the surface of the workbench and tapped his pencil. Sandy held her breath.

Finally, Mr. Whyte looked at Sandy and said, "Okay."

Sandy let her breath out and smiled. "Thanks, Mr. Whyte. I won't let you down. I promise."

"I know you won't, Sandy. Come on. We better get moving if we expect to get some work done."

Mr. Conrad and Mr. Brown met Sandy and Mr. Whyte at the shop door. After exchanging greetings, Mr. Whyte told the two men that Sandy would be the plumbing crew chief.

"Nice going, Sandy," Mr. Brown said, extending his hand to her. "Congratulations."

Sandy, still smiling, shook Mr. Brown's hand. "Thank you, sir. I'm really excited. I can hardly wait to get to work."

"I'm sure you'll do a good job," Mr. Brown said. He turned to Mr. Whyte. "The house looks great." Everyone turned to look at the building site. The walls were up, the roof was on, and it really looked like a house. Mr. Brown continued, "My wife is back from Boston, so she'll be out today. She's thinking of relocating the kitchen sink."

Mr. Whyte turned to Sandy and said, "Make sure you talk to Mrs. Brown today."

Sandy nodded and Mr. Conrad said, "Come on, Sandy. We have work to do, and Mr. Brown wants to talk to Mr. Whyte. Let's go tell the others your good news."

When Sandy and Mr. Conrad reached the house site, Mr. Conrad called the students together and announced that Sandy would be the plumbing crew chief.

"Not a moment too soon, Sandy," Al said, holding a piece of white PVC pipe with 90-degree **fittings** on each end. "A couple of the joints in that outside waterline didn't set up right and leaked, so Mr. Conrad told me to replace it. I measured that pipe three times and cut the new PVC real carefully, and it still doesn't fit! It's too long. What did I do wrong?"

"Looks like a good time to review some basic plumbing, Sandy. Are you up for it?" Mr. Conrad asked.

Sandy smiled. "Yes, sir." She turned to Al. "I'll bet you didn't adjust the length of the pipe for the fittings, Al."

"You bet right. I didn't adjust the length of the pipe for anything," Al told her.

Sandy took the length of pipe from Al and laid her tape measure across it. "When I measure this center to center I get 15½ inches, but when I measure it face to face I get 12 inches. Were you trying for 12 inches?"

"Yeah, I was," Al answered. "But what's this face to face stuff?"

"That describes part of the fitting." Sandy rummaged through a white box and pulled three different PVC fittings out. "The openings on these are the face—that's where the water flows through the fitting. The other part is the back. This edge is called the throat, but that name's hardly ever used. You measure the fittings on center." Sandy laid her tape measure across the center of each fitting vertically, horizontally, and diagonally.

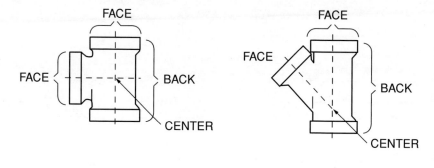

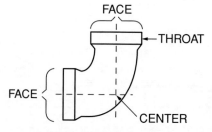

"The PVC pipe fits into the face of the fitting with an overlap, so they can be glued together. These fittings add length to a run of pipe," Sandy went on. "You need to adjust the pipe length so the section fits correctly. The fitting length is called the fitting allowance. Do you still have the old piece of PVC you cut out?"

"Right here." Al handed Sandy a length of PVC.

"When I measure the old section of PVC center to center, I get 12 inches, but I need to subtract the fitting allowance to get the correct length for the pipe."

"Wait, Sandy," Travis interrupted. "I'm getting lost. What is center to center?"

"It's just a way to measure pipe. I can measure center to center, back to back, face to face, and so on." Sandy moved her tape measure across the length of the pipe to demonstrate each method.

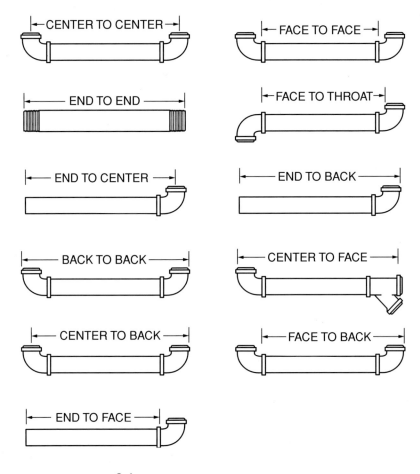

"On PVC pipe, I measure center to center because the manufacturer labels each box of fittings with the adjustment value." Sandy pointed to a picture on the side of a box. "Notice that the adjustment is given from the center of the fitting to the beginning of the throat. That's because the pipe slides all the way into the throat, overlapping it."

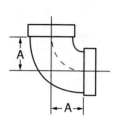

SIZE	BEND (SANITARY 90° TEE) ALL HUB		
	A	B	C
1½"	1¾"		
2"	2⁵⁄₁₆"		
3"	3¹⁄₁₆"		
4"	3⅞"		
6"	5"		
8"	6"		

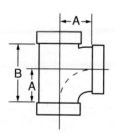

SIZE	SANITARY TEE ALL HUB		
	A	B	C
1½"	1¾"	2¾"	
2"	2⁵⁄₁₆"	3¹¹⁄₁₆"	
3"	3¹⁄₁₆"	4⅞"	
4"	3⅞"	6⅛"	
6"	5"	8½"	
8"	6"	10½"	

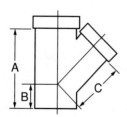

SIZE	WYE STREET (45° WYE) SPIGOT × HUB × HUB		
	A	B	C
1½"	4¾"	1⅞"	2⅞"
2"	5⅞"	2¼"	3⅝"
3"	8⅛"	3⅛"	5"
4"	10"	3⅝"	6⅜"

Key To Understanding

Remember, whatever you do to one side of an equation, you need to do to the other. Sandy subtracted 3½ from both sides of the equation to find the unknown, which was pipe length.

"This first one is a 90-degree bend. Since the PVC we're using is 1½ inches in diameter, the fitting allowance is 1¾ inches. I have two fittings so together the allowance is 3½ inches. I subtract the allowance for both fittings from the length of the whole section, which is 12 inches, and I get 8½ inches for my pipe length." Sandy wrote some figures in her notebook.

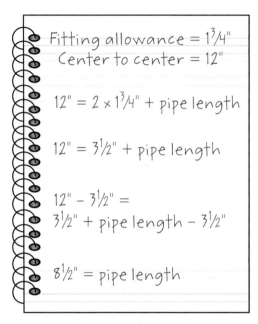

Fitting allowance = 1¾"
Center to center = 12"

12" = 2 × 1¾" + pipe length

12" = 3½" + pipe length

12" − 3½" =
3½" + pipe length − 3½"

8½" = pipe length

Sandy continued. "If you measure the old section of pipe face to face, you would need to add the length of the throat to the face-to-face value to get the correct length of pipe."

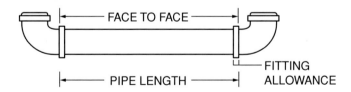

Practice Problems 9-1

Al needs to fabricate the pipe joints shown in the following diagram. Calculate the pipe length for the fittings specified in Questions 1 and 2.

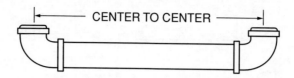

1. The center-to-center measurement of the entire pipe joint is 12 inches. Each fitting has a 2½-inch fitting allowance. The length of the pipe between the fittings is **7 inches**.

   12" − 2 × 2½" = 7"

2. The center-to-center measurement of the entire pipe joint is 15 inches. Each fitting has a 5-inch fitting allowance. The length of the pipe between the fittings is **5 inches**.

   15" − 2 × 5" = 5"

Travis needs to fabricate the pipe joints shown in the following diagram. Calculate the pipe lengths for the fittings specified in Questions 3 and 4.

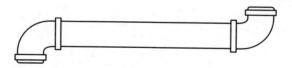

3. The center-to-center measurement of the entire pipe joint is 18 inches. Each fitting has a 2½-inch fitting allowance. The length of the pipe between the fittings is **13 inches**.

   18" − 2 × 2½" = 13"

4. The center-to-center measurement of the entire pipe joint is 15 inches. Each fitting has a 3⅞-inch fitting allowance. The length of the pipe between the fittings is **7¼ inches**.

   15" − 2 × 3⅞" = 7¼"

Tom needs to fabricate the pipe joints shown in the following diagram. Calculate the pipe length for the fittings specified in Questions 5 and 6.

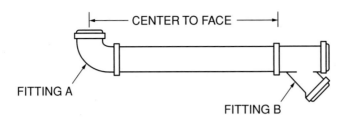

5. The center-to-face measurement of the entire pipe joint is 18 inches. (Hint: Notice that this measurement is not center to center.) Fitting A has a 1¾-inch fitting allowance. Fitting B has a 1⅞-inch fitting allowance. The length of the pipe between the fittings is **16¼ inches**.

   18" − 1¾" = 16¼"

6. The center-to-face measurement of the entire pipe joint is 12 inches. (Hint: Notice that this measurement is not center to center.) Fitting A has a 6-inch fitting allowance. Fitting B has a 3⅝-inch fitting allowance. The length of the pipe between the fittings is **6 inches**.

   12" − 6" = 6"

Later that morning, Mrs. Brown joined Mr. Brown at the house site. She drove Mr. Brown's old red pickup truck to the site. Mr. and Mrs. Brown stood by the truck, talking. Al stopped by and introduced himself.

"Yes, sir," Al said. "We're practically neighbors. I live with my grandmother on Sandy Lane. It's not far from here. I notice your truck is for sale. How much are you asking for it?"

"$1,500," Mr. Brown responded.

"Mmmm," Al said. "A little more than I can spend right now, but I'll keep it in mind."

Al escorted Mrs. Brown into the house and introduced her to Mr. Conrad. Mrs. Brown asked Mr. Conrad if she could move the location of the kitchen sink. Her kitchen cabinets were being custom made from walnut and the cabinetmaker told her he would waste less wood if the sink was moved one foot to the right. Mr. Conrad didn't think it was a good idea, but he called Sandy over to talk to Mrs. Brown.

"A foot's not a problem for the waterlines, Mrs. Brown," Sandy said. "We can offset them, but I need to talk to Mr. Whyte about moving the drain pipe."

"What's offset mean?" Mrs. Brown asked.

"When a pipe changes direction, it's called offset." Sandy drew in her notebook. "We'd have to move the hot and cold water lines to the right when we move the sink to the right."

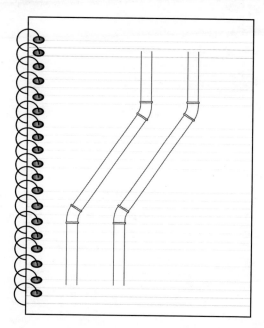

"Is it hard to figure out the angle of the offset?" Mrs. Brown asked.

"No, this would be a 45-degree offset. It's a real common offset in plumbing. The trick is to get the length of the pipe correct."

"Do you have time to explain it to me, Sandy?" Mrs. Brown asked. "I love puzzles."

Sandy smiled. "It's like this. An offset forms a right triangle. In this case, it's a 45-degree offset and the offset distance is 12 inches or one foot."

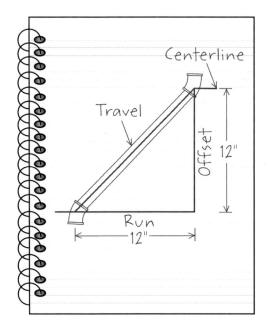

"As soon as I say a right triangle with a 45-degree angle, I know that the third angle of the triangle is 45 degrees too, because the sum of the angles of any triangle must be 180 degrees."

 **Math Speak**

A triangle is a closed shape that has three sides and three angles. The sum of all three angles must equal 180 degrees. A right triangle must have one angle that is 90 degrees, so the sum of the other two angles must be 90 degrees. 90 degrees + 90 degrees = 180 degrees.

When one angle of a right triangle is 45 degrees, the other angle must be 45 degrees, because 45 degrees + 45 degrees + 90 degrees = 180 degrees.

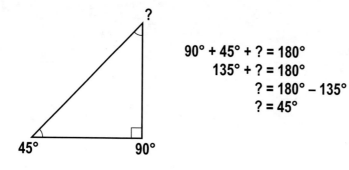

When one angle of a right triangle is 50 degrees, the other angle must be 40 degrees, because 50 degrees + 40 degrees + 90 degrees = 180 degrees.

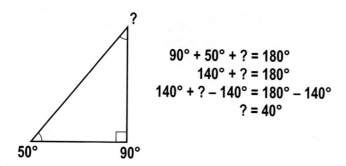

Practice Problems 9-2

Calculate the number of degrees in the third angle of each of the right triangles.

1.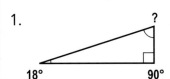

   $90° + 18° + ? = 180°$
   $108° + ? = 180°$
   $108° + ? - 108° = 180° - 108°$
   $? = 72°$

2.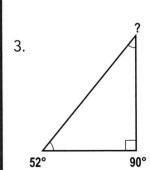

   $90° + 49° + ? = 180°$
   $139° + ? = 180°$
   $139° + ? - 139° = 180° - 139°$
   $? = 41°$

3.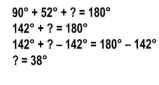

   $90° + 52° + ? = 180°$
   $142° + ? = 180°$
   $142° + ? - 142° = 180° - 142°$
   $? = 38°$

4.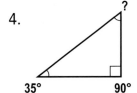

   $90° + 35° + ? = 180°$
   $125° + ? = 180°$
   $125° + ? - 125° = 180° - 125°$
   $? = 55°$

5.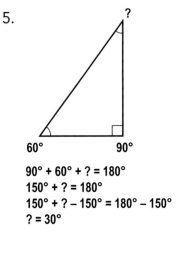

   $90° + 60° + ? = 180°$
   $150° + ? = 180°$
   $150° + ? - 150° = 180° - 150°$
   $? = 30°$

Sandy continued. "The 45-degree offset is so commonly used in plumbing that I know two sides are the same length—that's always true when you have a right triangle where the other two angles are both 45 degrees. In plumbing these sides are called run and offset, but in math these sides are labeled with the letters a and b."

Sandy drew a triangle. "The run and offset are both 12 inches, so sides a and b of the triangle are 12 inches, too. The third line is called travel. In math, this side of the triangle is labeled with the letter c and is called the hypotenuse. I know that I can figure out the length of line c using the Pythagorean theorem. I'm going to use my calculator to get the square root of 288 ($12^2 + 12^2$), which is 16.97 inches. I can subtract the fitting allowance and get the actual length I need to cut the pipe."

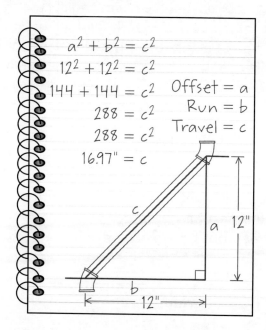

"That's pretty good, Sandy. I had no idea that the Pythagorean theorem was so useful," Mrs. Brown said.

"This brings up a question, Sandy. I work at a hospital, and there are these huge drainpipes running down the walls. Sometimes they have an offset to get around one thing or another. I can see how you could measure the run, but the offset is from floor to ceiling. Do you need to estimate it so you can calculate the travel?"

"No. In that case, I would use a constant. See, these two charts show the constants for plumbing angles." Sandy showed her a table from a plumbing catalog.

Finding Travel When Run is Known		Finding Travel When Offset is Known	
Angled Fitting	Constant	Angled Fitting	Constant
11¼°	1.019	11¼°	5.126
22½°	1.082	22½°	2.613
45°	1.414	45°	1.414
60°	2.0	60°	1.1547
72°	3.236	72°	1.05

"When I know the run and the offset angle, I use the first table to get the constant." Sandy said. "I just multiply the run by the constant, and I get the travel. When I know the offset angle and the run, I use this other table to find my constant and multiply it by the run to get the travel."

Sandy drew a plumbing joint in her notebook. "Here's an example. This offset is 60 degrees. I know the run is 12 inches, so the constant is 2. Then I multiply the run of 12 inches by 2 and I get 24 inches for the travel. I don't need to do this next part because I know that the calculation is correct, but I'm doing it to show you that you can trust the constants."

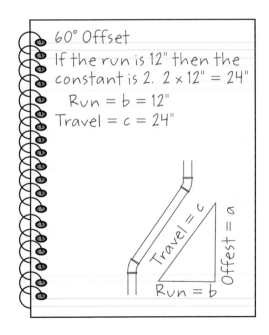

Sandy went on. "I use the Pythagorean theorem to solve for the offset, and I get 20.79 inches. Then I look up the constant for a 60-degree fitting when I know the offset, and it's 1.1547. So I multiply 1.1547 by 20.79 inches, and I get 24 inches for the travel. See, the constants work!"

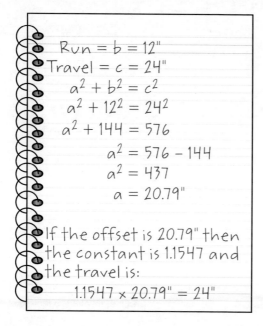

"Wait a minute, Sandy. Do I recognize these constants?" Mrs. Brown asked.

Sandy smiled. "You probably do, Mrs. Brown. They're ratios of the lengths of the sides of the triangle."

"Isn't that something," Mrs. Brown said. "I always did very well in math class, but this is the first time I ever saw a use for some of it!"

Math Speak

The Pythagorean theorem is used for right triangles. Using the Pythagorean theorem, you can calculate the length of the third side if you know the length of two sides.

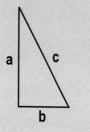

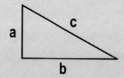

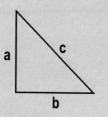

Know a and b, calculate c:

$a^2 + b^2 = c^2$

$\sqrt{a^2 + b^2} = \sqrt{c^2}$

$\sqrt{a^2 + b^2} = c$

Key To Understanding

A ratio is a comparison between two numbers. It can be written as a fraction (½), with a colon (1:2), or with the word "to" (1 to 2). This comparison is shorthand to say for every one of the first number, I have 2 of the second number. For example, if the ratio of x/z is ½, then for every x, there are 2z. If x = 12, then z = 24. If the ratio of x/z is 2/1, which can be expressed as the single digit "2," then for every 2x, there is 1z. If x = 12, then z = 6.

In a 45-degree right triangle, the ratio of sides a and b is 1 to 1, which means that both sides are equal in length. This is always true in a 45-degree right triangle. If side a is 1 inch, then side b is 1 inch. If side a is 12 inches, side b is 12 inches. If side a is 4 inches, side b is 4 inches.

Here's an example of ratios at work for rectangles:

    L:W = 3:1
    if L = 3 then W = 1
    if L = 6 then W = 2
    if L = 1 then W = ⅓

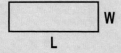

Mind Games

Sandy said that in a 45-degree right triangle, the lengths of sides a and b must be equal. Let's try it!

First, draw a 45-degree angle with a base of 3 inches.

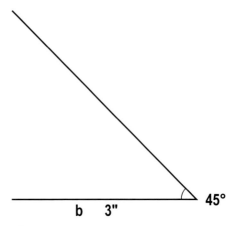

Next, close the angle so that you have a right triangle, and then measure the length of side a. If you were very precise, it will be 3 inches.

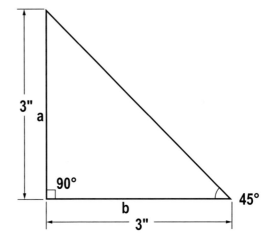

Now, draw a smaller right triangle inside the larger triangle. Make the base 2 inches, and then measure the length of side a of the smaller triangle. If you were very precise, it will be 2 inches.

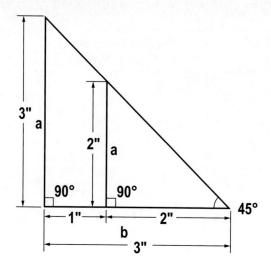

Try one more. This time use a base of 1 inch and measure the length of side a. You guessed it! It will be 1 inch. So the ratio of sides a and b of a 45-degree right triangle is 1:1.

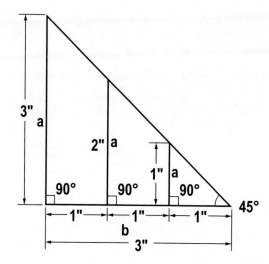

Practice Problems 9-3

Using the table shown below, complete the following problems.

Finding Travel When Run is Known		Finding Travel When Offset is Known	
Angled Fitting	Constant	Angled Fitting	Constant
11¼°	1.019	11¼°	5.126
22½°	1.082	22½°	2.613
45°	1.414	45°	1.414
60°	2.0	60°	1.1547
72°	3.236	72°	1.05

1. The fitting angle is 45 degrees, and the run is 12 feet. What is the travel?

   45° constant = 1.414
   Run = 12'
   1.414 × 12' = 16.968'

2. The fitting angle is 11¼ degrees, and the run is 21 feet. What is the travel?

   11¼° constant = 1.019
   Run = 21'
   1.019 × 21' = 21.399'

3. The fitting angle is 60 degrees, and the offset is 14 feet. What is the travel?

   60° constant = 1.1547
   Offset = 14'
   1.1547 × 14' = 16.1658'

4. The fitting angle is 72 degrees, and the run is 57 feet. What is the travel?

   72° constant = 3.236
   Run = 57'
   3.236 × 57' = 184.452'

5. The fitting angle is 22½ degrees, and the offset is 16 feet. What is the travel?

   22½° constant = 2.613
   Offset = 16'
   2.613 × 16' = 41.808'

Later that week, two men brought a backhoe to the building site and dug a hole on the north side of the house. Sandy knew this was the wrong location because the ground was flat on that side of the house. The land at the back of the house sloped gently away from the house, so she had decided that was the best place for the septic tank and that was where the **perk test** was performed. Sandy tried to stop the workers from digging the hole, but they were Hispanic and didn't understand English very well.

"Stop," Sandy shouted in Spanish, waving her arms. "Jorge, get over here!"

Jorge was in the house gluing PVC joints with Phil. Both came running out of the house.

"Hey, Sandy, what's happening?" Jorge asked as he stared at the deep hole next to the house.

"These guys don't speak English, and they didn't teach me how to say septic tank in Spanish class. I need you to tell them the septic tank goes over at that marker." Sandy pointed to an orange flag stuck in the ground behind the house.

"No problem," Jorge said as he jogged over to the backhoe and engaged the two workers in a rapid exchange of Spanish. The man operating the backhoe nodded, while the man on the ground walked toward the flag.

"He's going to fill in this hole and then dig one out back," Jorge told Sandy and Phil when he rejoined them. "They told me the inspector from the health department is on his way. They just finished a job with him in Gainesville." In most areas, septic tank installations have to be inspected by the health department instead of codes enforcement, because the risk of contaminating the water supply is so great.

Key To Understanding

A grade is the slope of a line in reference to a horizontal line.

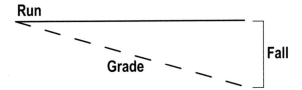

The horizontal line is called the run. The sloped line is called the grade. The vertical distance between the end of the run and grade lines is called the fall. Grade is a ratio between the fall and the run; it's calculated like this:

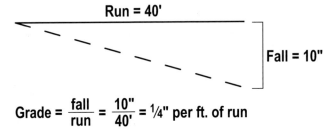

$$\text{Grade} = \frac{\text{fall}}{\text{run}} = \frac{10"}{40'} = 1/4" \text{ per ft. of run}$$

Notice how the grade is given in inches per foot. This means that for every one foot of run, the grade decreases the specified inches. The drawing shows you that the ground drops 10 inches for every 40 feet of horizontal run. The grade tells you the ground drops ¼ inch every 1 foot of horizontal run. Plumbers frequently use grades from ⅛ inch to ½ inch per foot.

"Good," Sandy said. "While they're doing that, we had better plan the grade of the drain pipe from the house to the septic tank. Let's go back into the house and get out of this sun. I'm roasting!"

"Why is the grade of a drainpipe so important, Sandy?" Phil asked. "We don't worry about slope in water pipes."

"Because the water in water pipes is pressurized," Sandy answered. "Drain pipes need to work with gravity. Installing a drain pipe on a slope helps it drain."

Key To Understanding

You know how sometimes you turn on a water faucet and water gushes out with unexpected force? That water is under a lot of pressure, and pressure forces the water to pour out of the faucet. Other times, you turn on a faucet, and the water trickles out in a slow stream. That water is under very little pressure.

There are lots of ways to pressurize water, but the oldest method is by elevation. For every foot of elevation, there is an increase in water pressure of 0.433 **pounds per square inch (psi)**. That's why water is often stored in elevated water tanks. The higher above the outlet the water is stored, the greater the water pressure.

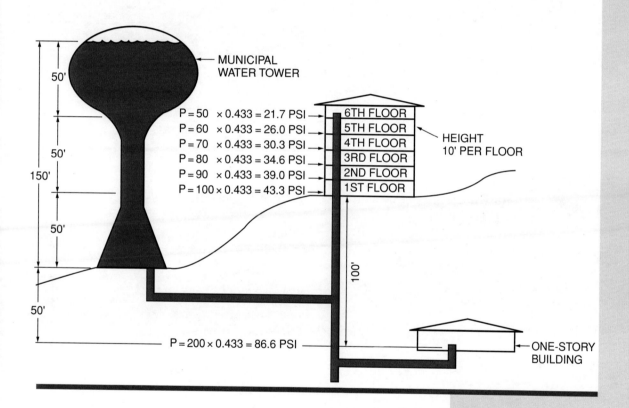

Sandy, Phil, and Jorge calculated the fall for the drain-pipe trench. They knew that the grade was supposed to be ¼ inch per foot, and they knew the septic tank was going to be 40 feet from the house.

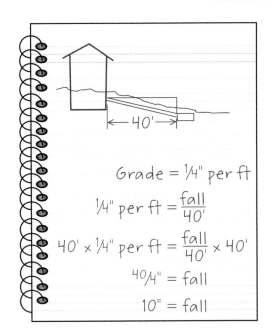

$$\text{Grade} = \tfrac{1}{4}\text{" per ft}$$
$$\tfrac{1}{4}\text{" per ft} = \frac{\text{fall}}{40'}$$
$$40' \times \tfrac{1}{4}\text{" per ft} = \frac{\text{fall}}{40'} \times 40'$$
$$\tfrac{40}{4}\text{"} = \text{fall}$$
$$10\text{"} = \text{fall}$$

Practice Problems 9-4

1. Draw a diagram that shows a grade of ½ inch per foot with a run of 5 feet.

   Run = 5'
   Grade = ½" per foot

   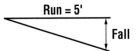

2. Calculate the fall at 5 feet for the grade described in Question 1.

   ½" per foot × 5' = 2½"

3. Calculate the grade of a drainpipe with a run of 18 feet and a fall of 1½ feet.

   Run = 18' Fall = 1½', so the grade = 1½' ÷ 18' = 18" ÷ 18' = 1" per foot

4. The grade of a drainpipe is ¼ inch per foot. Calculate the fall at 80 feet.

   Grade = ¼" per foot
   At 80 feet, the fall = 80' × ¼" per foot = 20"

Mr. Conrad found the students sitting on the floor of what would be the Browns' dining room. "Have you seen Tom?" he asked. "I haven't seen that boy since lunch."

"I think he's at the shop with Mr. Brown and Mr. Whyte," Jorge told him.

"Thanks." Mr. Conrad looked at the figures Sandy had written in her notebook. "Are you planning the drainpipe installation? This looks good. Have you planned how you're going to check the trench grade?"

Sandy looked perplexed. "What do you mean, Mr. Conrad? Can't I just measure the depth of the trench?"

Mr. Conrad shook his head. "No, Sandy. This is the fall and this is the run." Mr. Conrad drew in his notebook. "See how the ground is at different levels? Measuring the depth of the trench isn't going to help you."

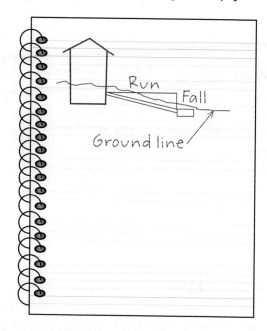

Sandy, Phil, and Jorge studied the diagram Mr. Conrad had drawn.

"You guys think on it some," Mr. Conrad told them as he handed the diagram he had drawn to Sandy. "I'll be in the garage. Al and I are planning the heating and air conditioning installation. If you can't figure it out, call me."

Sandy, Phil, and Jorge looked at the drawing, and then went outside to look at the ground. Sure enough, the ground sloped down toward the septic tank. The students discussed various ways to measure the correct depth of the trench accurately, but they discarded all of their ideas.

"Hi, guys. What's going on?" Tom asked, walking toward them. He was carrying the big 100-foot measuring tape and a bunch of markers. He was headed out to mark the trench for the electrical cable to the house.

Sandy told Tom the problem. Tom looked at the diagram and looked at the trench. None of the students noticed Mr. Conrad standing on the nearby porch, listening.

"This is just like that thing Mr. Burke showed us during the survey," Tom said. "That thing called breaking the tape. Jorge, you hold this tape at the drainpipe coming from the house. Phil, you take the tape all the way out to the septic tank. I'll put my line level on the tape to make sure it's level. Then Sandy can measure from the tape to the trench floor to check the fall."

"I don't get it," Jorge said.

"I don't either," Sandy said.

Tom drew in his notebook. "If I take a tape measure and run it out 40 feet and then make sure it's level, I can measure from the tape to the bottom of the trench and check the grade. With a grade of ¼ inch per foot, I expect the trench floor to be 2.5 inches below the tape at 10 feet, 5 inches at 20 feet, 7.5 inches at 30 feet, and finally 10 inches at 40 feet."

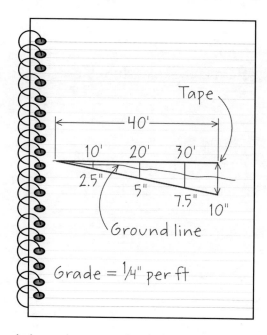

Sandy and the others studied the diagram, then Sandy smiled. "That's pretty neat, Tom."

Tom looked up and saw his father standing on the porch. Mr. Conrad looked at Tom for a long moment. He gave Tom a small smile, and then turned and walked into the house.

Chapter 9 Review Problems

Calculate the number of degrees in the third angle of each of the right triangles for Questions 1 through 3.

1.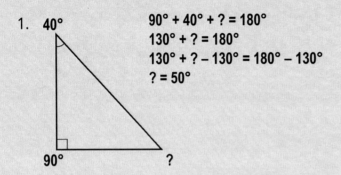

   $90° + 40° + ? = 180°$
   $130° + ? = 180°$
   $130° + ? - 130° = 180° - 130°$
   $? = 50°$

2.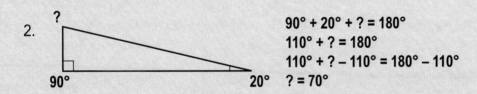

   $90° + 20° + ? = 180°$
   $110° + ? = 180°$
   $110° + ? - 110° = 180° - 110°$
   $? = 70°$

3.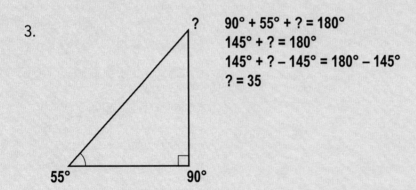

   $90° + 55° + ? = 180°$
   $145° + ? = 180°$
   $145° + ? - 145° = 180° - 145°$
   $? = 35$

Using the table shown below, complete Questions 4 through 6.

Finding Travel When Run is Known		Finding Travel When Offset is Known	
Angled Fitting	Constant	Angled Fitting	Constant
11¼°	1.019	11¼°	5.126
22½°	1.082	22½°	2.613
45°	1.414	45°	1.414
60°	2.0	60°	1.1547
72°	3.236	72°	1.05

4. The fitting angle is 45 degrees, and the run is 71 feet. What is the travel?

   45° constant = 1.414
   Run = 71'
   1.414 × 71' = 100.394'

5. The fitting angle is 11¼ degrees, and the offset is 21 feet. What is the travel?

   11¼° constant = 5.126
   Offset = 21'
   5.126 × 21' = 107.646'

6. The fitting angle is 60 degrees, and the run is 18 feet. What is the travel?

   60° constant = 2
   Run = 18'
   2 × 18' = 36'

7. Calculate the grade of a drainpipe with a run of 24 feet and a fall of 1 foot.

   Grade = fall/run
   Grade = 12"/24' = ½" per foot

8. Draw a diagram that shows a grade of ¼ inch per foot with a run of 12 feet and calculate the fall at 12 feet.

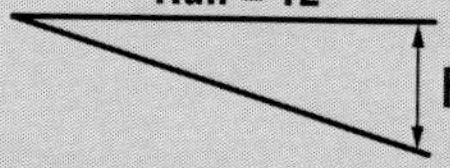

   Run = 12'
   Fall

   Fall = 12' × ¼" per foot = 12/4 = 3"

9. Using the following diagram, calculate the distance between the tape and the trench floor at 5 feet.

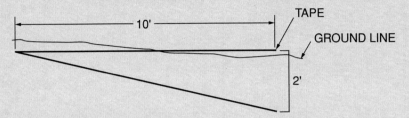

Grade = fall/run
Grade = 24"/10'
Grade = 2⅖" per foot, so at 5' the fall = 2⅖" × 5 = 10¹⁰⁄₅ = 12"

10. Write a short essay describing how Tom, Sandy, and the other students check the grade of the drainage pipe trench.

Answers will vary. Have one of the students write their answer on the board and discuss it.

Careers in Construction—Plumber

Plumbers design the piping systems that distribute water and remove waste from buildings. Plumbers need to know about local codes, as well as mathematics and physics. They also need to be good with their hands, because plumbers sometimes need to do welding and soldering. In most parts of the United States, plumbers need to be licensed. It can take four to five years of training to become a licensed plumber. During most of the training time, you would work as a plumber's helper and get paid, but part of the time needs to be spent in study—either in a classroom or on your own. As you can tell from this chapter, plumbers use a lot of math. If you want to know more about plumbing, check at your local community college or look on the National Center for Construction Education and Research website (www.NCCER.org).

History—Plumbing

London plumber Thomas Crapper has been credited with the invention of the self-contained flush toilet. Crapper was responsible for many improvements in plumbing. During the late 19th century, he was granted patents for everything from enhancements to drains to manhole covers to pipe joints. Around the same time, the first "modern" toilet appeared; it was a U-bend siphoning system to flush the pan where the waste was deposited. Records show that the patent for this device was actually issued to Albert Giblin. It is likely that Crapper bought the patent rights from Giblin, and then marketed the device himself. Crapper's association with the toilet may have originated during World War I, when soldiers used his name as a synonym for the toilets they saw as they passed through England to the front lines in Europe.

Terms

Fitting: In plumbing, a general term that describes valves, joints, faucets, and other plumbing fixtures.

Perk test: A test to determine how fast water passes through soil.

Pounds per square inch (psi): A measurement of how much pressure is applied on one square inch of surface. If you weigh 100 pounds and you stand on one 1-inch block of wood so that one square inch of the wood is in contact with the ground, you are exerting 100 pounds of pressure on the ground under the block. If you stand on two 1-inch blocks of wood, you are exerting 50 pounds of pressure on the ground under each block.

NOTES

Chapter 10
Shocking, Simply Shocking

This chapter discusses angles. Most people understand a circle is 360°, but you can review this in a fun way by relating it to aviation. Pilots use azimuth for direction and runways are marked with two digits that denote the approach azimuth (magnetic) direction. The designation is always rounded to the nearest tenth, so 59° is designated 06, while 159° is designated 16. On the board, draw a circle and mark the 0/360° point as north, then draw a runway over the circle and explain the angle of approach.

NOTES

Chapter 10

Shocking, Simply Shocking

Mr. Brown was pleased with everyone's work, but especially Tom's. Tom had called the rural electric company that supplied electricity to the Browns' house and found that the electric company would pay for the input power cable all the way up to the meter. If the Browns put one meter on the house and a second one on the shop, it would save them thousands of dollars in cabling costs. Mr. Brown was so pleased he asked Mr. Whyte if the students could wire his shop in addition to the house.

Tom laid a piece of metal conduit on the workbench. "In the shop, we're going to use metallic tubing as electrical conduit. It's abbreviated EMT for electrical metallic tubing. We'll use a conduit bender to bend the tube. When you make a 90-degree bend in conduit, you're making part of a circle." He scribbled in his notebook and showed his classmates.

Tom went on. "This is a 90-degree bend because it's ¼ of the circumference of a circle, and ¼ of a circle is 90 degrees."

"Hey, Tom," Al said. "I don't remember about circles, can you give me a quick review?"

Tom nodded. He thought for a moment and then flipped to a blank page of his notebook and drew a circle. "A circle is a closed curved line. The distance around the circle is called the circumference—this is like the perimeter in a square, rectangle, or triangle. All of the points on the line are the same distance from the center of the circle, and the distance from the center to any point on the line is called the radius. The distance from one side of the circle through the center to the opposite side is called the diameter. There are 360 degrees in a circle, and when a circle is divided into four equal parts, they're called quadrants."

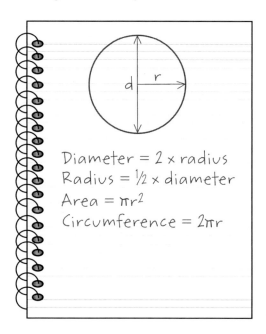

Diameter = 2 × radius
Radius = ½ × diameter
Area = πr^2
Circumference = $2\pi r$

Tom flipped the page of his notebook back to the previous drawing and pointed. "The distance between points A and B—that's called an arc. And in this case, it's ¼ of the circumference of the circle. Since this bend takes up 90 degrees of the circle, it's called a 90-degree bend and is called the take-up—meaning the length of conduit that's used to make the bend."

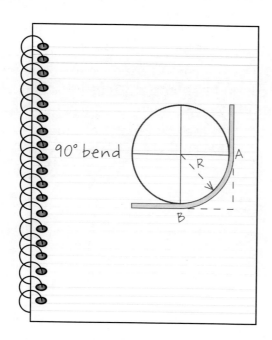

 Math Speak

- The distance from the center to any point on the curved line is always the same. This distance is called the radius and is abbreviated with the letter r. The plural for radius is radii.

- All points on a circle are the same distance from the center.

- The shortest distance from any point on the curve through the center to a point directly opposite is called the diameter. Diameter is abbreviated with the letter d. The diameter is equal to twice the radius (d = 2r).

- A circle can be divided into 360 parts. Each part is called a degree. Therefore, one degree = 1/360 of a circle, and a circle is 360 degrees.

- The distance around the outside of the circle is called the circumference. It can be calculated with the equation: circumference = πd, where π is a constant approximately equal to 3.14 and d is the diameter. Since the diameter of a circle is two times the radius, the circumference can also be calculated with the formula π2r or 2πr.

"The take-up length depends on the diameter of the conduit," Tom went on. "Since the 90-degree bend is one of the most commonly used when you're installing electrical conduit, it's easy to find information about take-up lengths so you don't need to figure them out. Here's a table that the electrician gave me." He showed them a simple table he had copied into his notebook.

EMT Diameter	Take-Up
1/2"	5"
3/4"	6"
1"	8"
1 1/4"	11"

"Sometimes a 90-degree bend is too much. You just need to make a little bend to get around something or to get into the junction box. In that case, you make a bend called an offset," Tom said.

Jorge straightened up and asked, "Hey, is this like an offset in plumbing? The one where you use a constant to figure out the travel?"

"Almost," Tom said.

"This is great. We already know about plumbing offsets and constants, so electrical conduit offsets will be a piece of cake," Al exclaimed.

"Well, it's a little different," Tom said with a shake of his head. "In plumbing, offsets are limited to the angle of the fittings. In electrical, the bends can be anything between 0 and 90 degrees, so you may need to use some math to figure out the constant."

"I might have guessed," Al said.

Mr. Whyte laughed. "You didn't think you'd get out of doing math, did you, Al?"

"I can hope, Mr. Whyte. I can hope."

Mr. Conrad smiled, and then turned to his son. "Continue, Tom. I've been bending conduit since I can't remember when, but I'm not sure I ever knew the math behind it."

Tom stared at his father for a few seconds, then continued. "Let's say we want to make a couple of little bends to get around something." Tom's voice trembled as he began but it became steadier and firmer as he continued. "You need to know where to place the first bend, and then the distance between the first and the second bend. When you're bending conduit and make two bends that are between 0 and 90 degrees, you get a nice neat right triangle. The hypotenuse of the triangle is the length between the bends, and side a of the triangle is called the offset depth."

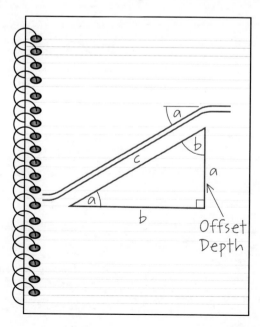

 Math Speak

The Pythagorean theorem is used for right triangles. Using the Pythagorean theorem, you can calculate the length of the third side if you know the length of two sides, like this:

$$a^2 + b^2 = c^2$$

Side a is the rising side. Remember it with the word altitude. Side b is the base. Side c is always the long side of the triangle.

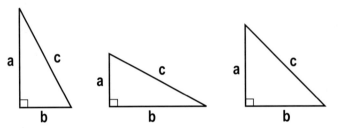

The sum of the three angles in any triangle must equal 180 degrees. A right triangle has one 90-degree angle, so that means the sum of the other two angles must equal 90 degrees.

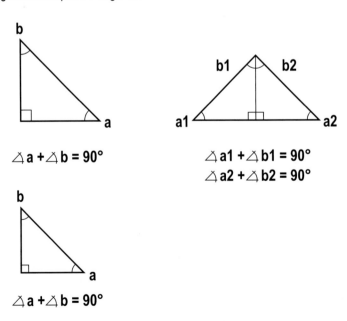

$\triangle a + \triangle b = 90°$

$\triangle a1 + \triangle b1 = 90°$
$\triangle a2 + \triangle b2 = 90°$

$\triangle a + \triangle b = 90°$

Practice Problems 10-1

Complete the following problems.

1. A circle has a diameter of 6 inches. What is the radius? What is the circumference?

   r = ½ × d = ½ × 6" = 3"; c = πd = 3.14 × 6" = 18.84"

2. A circle has a radius of 7 inches. What is the diameter? What is the circumference?

   d = 2r = 2 × 7" = 14"; c = πd = 3.14 × 14" = 43.96"

3. A circle has a circumference of 15.7 inches. What is the radius?

   15.7" = 3.14 × d = 5"
   r = ⁵⁄₂ = 2½"

4. Calculate the size of the angles in the following figure.

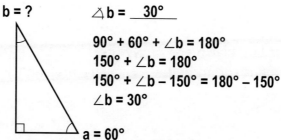

   △b = __30°__

   b = ?

   90° + 60° + ∠b = 180°
   150° + ∠b = 180°
   150° + ∠b − 150° = 180° − 150°
   ∠b = 30°

   a = 60°

5. Calculate the size of the angles in the following figure.

   △b = __58°__

   b = ?   90° + 32° + ∠b = 180°
   122° + ∠b = 180°
   122° + ∠b − 122° = 180° − 122°
   a = 32°   ∠b = 58°

6. Calculate the size of the angles in the following figure.

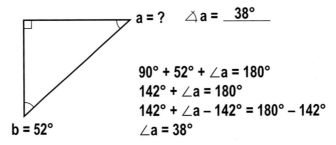

$a = ?$ $\triangle a = \underline{38°}$

$b = 52°$

$90° + 52° + \angle a = 180°$
$142° + \angle a = 180°$
$142° + \angle a - 142° = 180° - 142°$
$\angle a = 38°$

7. Calculate the size of the angles in the following figure.

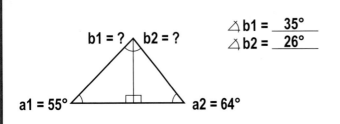

$b1 = ?$ $b2 = ?$
$a1 = 55°$ $a2 = 64°$

$\triangle b1 = \underline{35°}$
$\triangle b2 = \underline{26°}$

$90° + 55° + \angle b1 = 180°$
$145° + \angle b1 = 180°$
$145° + \angle b1 - 145° = 180° - 145°$
$\angle b1 = 35°$

$64° + 90° + \angle b2 = 180°$
$154° + \angle b2 = 180°$
$154° + \angle b2 - 154° = 180° - 154°$
$\angle b2 = 26°$

8. You need to cut a rectangular piece of plywood into three triangular shaped pieces as shown in the diagram below. Calculate the angles of the triangles.

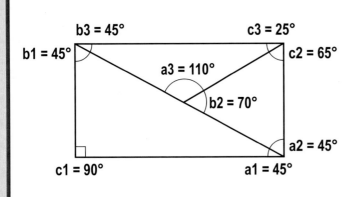

b3 = 45°   c3 = 25°
b1 = 45°   c2 = 65°
         a3 = 110°
              b2 = 70°
                    a2 = 45°
c1 = 90°   a1 = 45°

$45° + 90° + \angle a1 = 180°$
$135° + \angle a1 = 180°$
$135° + \angle a1 - 135° = 180° - 135°$
$\angle a1 = 45°$

$45° + \angle b3 = 90°$
$45° + \angle b3 - 45° = 90° - 45°$
$\angle b3 = 45°$

$45° + \angle a2 = 90°$
$45° + \angle a2 - 45° = 90° - 45°$
$\angle a2 = 45°$

$110° + 45° + \angle c3 = 180°$
$155° + \angle c3 = 180°$
$155° + \angle c3 - 155° = 180° - 155°$
$\angle c3 = 25°$

$\angle c2 + \angle c3 = 90°$
$\angle c2 + 25° = 90°$
$\angle c2 + 25° - 25° = 90° - 25°$
$\angle c2 = 65°$

$45° + \angle b2 + 65° = 180°$
$110° + \angle b2 = 180°$
$110° + \angle b2 - 110° = 180° - 110°$
$\angle b2 = 70°$

– 10.8 –

Hints:

Don't be overwhelmed by all of the missing information. Zero in on what you know.

a. The question says in part "…a rectangular piece of plywood…" A rectangle is a four-sided figure with four 90-degree angles, so you know that each corner equals 90 degrees.

b. Angle b1 is 45 degrees. If angle c1 is 90 degrees, then angle a1 must be **45 degrees** and angle a2 equals **45 degrees**.

c. If angle b1 is 45 degrees and the sum of angles b1 and b3 equals 90 degrees (because the angle of those two triangles started out as a 90-degree corner of a rectangle), then angle b3 must equal **45 degrees**.

d. If angle b3 equals **45 degrees** (from hint 3) and angle a3 equals 110 degrees and you know that the sum of the angles of all triangles must equal 180 degrees, then angle c3 equals **25 degrees**.

e. Use what you know and calculate the values of angles b2 and c2 on your own. Angle b2 equals **70 degrees** and angle c2 equals **65 degrees**.

"Remember how Mr. Whyte is always telling us that in a 45-degree right triangle, sides a and b are always equal?"

Al jumped up from his place on the floor. "Do I ever!" Al imitated Mr. Whyte twisting his mustache. "There are few things you can count on in life, but one of them is that the ratio between sides a and b of a 45-degree right triangle will always be 1 to 1."

Mr. Whyte was sitting on the workbench. He wagged his finger back and forth in a no motion. "Tsk, tsk. It's not nice to tease the teacher. It's too bad you're going to fail my class this year, Al."

"It's okay, Mr. Whyte," Al said. "I'd really like to spend another year with you."

Mr. Whyte's eyes opened wide. "Yikes. I didn't think of that!"

The group laughed and Tom continued. "It's just like Al said. When you compare the lengths of sides a and b of a 45-degree right triangle, their ratio is 1 to 1. Ratios are sometimes written as a fraction, a decimal, or a whole number."

Key To Understanding

A ratio is a comparison between two numbers. When you see a ratio, you know that it is comparing two numbers. With a 1 to 1 ratio (or 1:1), the two numbers are equal.

With a ratio of 1 to 2 (1:2), the second number is twice the first—you can also say the first number is ½ the second number. A 1:2 ratio is expressed in decimal as 0.5, so when someone says two numbers have a ratio of 0.5, you need to put a 1 after the 0.5 (0.5:1). This says that for every 0.5 you have of the first number, you have 1 of the second number. When the first number is 3, ask yourself "How many 0.5s are there in 3?" There are 6. In a 0.5 ratio when the first number is 3, the second number must be 6.

When you want to express the decimal ratio of 0.5 as a fraction, you place a 1 on top (that's the numerator). And then ask yourself how many 0.5s are there in 1? There are 2; that goes on the bottom (the denominator), so the fractional equivalent of the ratio 0.5 is ½.

When you want to express a ratio as a fraction, you need to work with whole numbers. When the ratio is 0.75, you need to ask yourself what is the smallest number that can be evenly divided by 0.75? It is 3, so 3 is the top number. Then you ask yourself, how many times does 0.75 go into 3? It's 4, so you express 0.75 as the fraction ¾.

You should recognize these calculations as simple decimal-to-fraction and fraction-to-decimal conversions. A ratio is just an easy way to express a relationship between two numbers so that when you know one of the numbers, you can calculate the other.

Practice Problems 10-2

Complete the following problems.

1. The ratio of x to y is 3:1.

    When x is 9, y is **3**.

    When x is 3, y is **1**.

    When x is 1, y is **⅓**.

2. In a right triangle, the ratio of side a to side b is 2:1.

    When side b is 2 inches, side a is **4"**.

    When side b is 2 inches, side c is **4.47"**.

    Hint: Draw a right triangle and label the sides. Use the Pythagorean theorem to find the length of side c.

    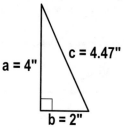

    $c^2 = a^2 + b^2$ $\sqrt{c^2} = \sqrt{20}$
    $c^2 = 2^2 + 4^2$ $c = 4.47"$
    $c^2 = 4 + 16$

3. A rectangle has a ratio of L:W = 1:3. Side L = 6 inches, so side W = **18 inches**. The area of the rectangle is **108 square inches**.

    SA = LW
    SA = 6" × 18"
    SA = 108 square inches

4. The ratio of any circle's radius to its diameter is 1:2.

    When a circle has a radius of 2, its diameter is **4**.

    When a circle has a radius of 1, its diameter is **2**.

    When a circle has a radius of 8, its diameter is **16**.

    When a circle has a diameter of 4, its radius is **2**.

    When a circle has a diameter of 7, its radius is **3½**.

    When a circle has a diameter of 8, its radius is **4**.

5. The ratio between two numbers is 0.25. That means the ratio is 0.25: **1** or 1:**4**.

0.25 = 0.25:1 so the first number is always ¼ of the second number and the second number is always 4 times the first number (1:4, 2:8, 3:12…)

6. The ratio of x:y is 0.75. The fractional equivalent is **¾**.

   When x = 3, y = **4**.

   When x = 6, y = **8**.

   When x = 9, y = **12**.

   When x = 1, y = **1⅓**.

   When x = 2, y = **2⅔**.

Tom went on. "Knowing that in a 45-degree right triangle the lengths of sides a and b are always equal tells me a lot. And now I'm going to relate what I know about a 45-degree right triangle to other right triangles."

Tom pulled out his notebook. "I'm going to draw a right triangle that has a base of 1¼ inches. Next, I'm going to use my protractor to make one 50-degree angle. Finally, I'll close the triangle and measure side a. It looks pretty close to 1½ inches. I'm going to convert the fractions to decimal, and I get 1.5 inches for side a and 1.25 inches for side b. See this squiggly line next to the 1.5 inches? That means approximately. It's really hard to measure accurately with a ruler, so I use the squiggly line next to the 1.5 inches to show that it's close, but not exact."

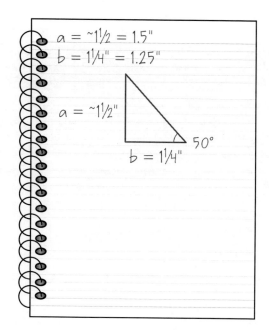

"Let me interrupt here, Tom," Mr. Whyte said. "You notice that Tom said the length of the line was close to 1.5 inches. If the length were closer to 1 7/16 inches than it was to 1½ inches, you would need to convert the 1 7/16 inches to decimal, which is 1.4375 inches. Don't get lazy and try to tell me, 'Oh, Mr. Whyte, I really, really thought the length of the line was closer to 1½ inches.'" Mr. Whyte did a perfect imitation of Al.

Al laughed and said, "You got me, Mr. Whyte."

"I know you want to use the nice easy 1.5 inches rather than taking the time and effort to convert the 7/16ths to a decimal. It's one of the oldest tricks in the book—I know. I tried to use it myself when I was your age."

"See, Mr. Whyte," Al said. "We got a lot in common."

Mr. Whyte rolled his eyes, slapped his palm to his forehead, and said, "Heaven help me."

The group erupted in laughter.

 **Mind Games** Sometimes in math, you need to be a bit of a detective. You start with what you know and use it to calculate more information. When you are working with a 45-degree right triangle, you know that the third angle must be 45 degrees too, because the sum of all angles must be 180. (The right angle is 90 degrees, the second angle is 45 degrees, so the third angle must be 45 degrees: 90 + 45 + 45 = 180.) The second thing you know is that in a 45-degree right triangle, the lengths of sides a and b are equal, so when you know the length of one side, you automatically know the length of the other side and by using the Pythagorean theorem, you can calculate the length of side c.

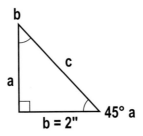

45° Right Triangle

If △a = 45°, then △b = 45° because the sum of a + b = 90°.

If b = 2", then a = 2" because the ratio of a:b is 1:1 in a 45° right triangle.

If a and b = 2", then $c = \sqrt{2^2 + 2^2}$

$$c = \sqrt{4 + 4}$$
$$c = \sqrt{8}$$
$$c = 2.83"$$

If a = b, then the ratio of a to b is 2:2, which is the same as 1:1, 3:3, 5:5, etc.

If a = 2" and c = 2.83", then the ratio of a to c is 2:2.83, which is the same as 1:1.41.

If a = 5", then b = 5" because the ratio of a:b is 1:1 and c = 7.05" because the ratio of a:c is 1:1.41 so when a = 5", c = 5" × 1.41 or 7.05".

Sandy went over 45-degree right triangles in Chapter 9. If you are confused, read Chapter 9 again, and then reread Chapter 10 to this point. Tom is going to move on to apply what you know about 45-degree right triangles to other triangles.

"Back to the problem with the 50-degree right triangle," Tom said. "Since I know the lengths of sides a and b, I can calculate the length of side c with the Pythagorean theorem. I get 1.96 inches."

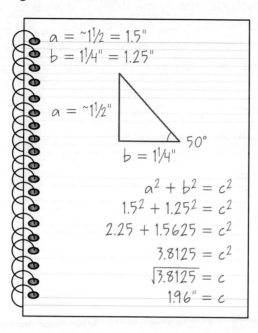

"I can take a ratio of side c to side a and I get 1.3."

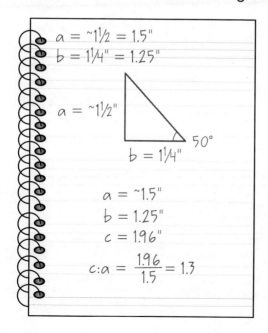

"Tom," Phil interrupted. "Where the heck is this going?"

"The lengths of the sides of a right triangle are related to the angles and they are constants. For every right triangle with a 50-degree angle here," Tom said, pointing to the 50-degree angle on his diagram. "I can say that the length of line a times the ratio of lines c to a equals the length of line c. See?" Tom scribbled in his notebook.

"Tom," Phil said. "You're going in a circle. You just calculated the length of side c. You don't need a ratio to find it again."

"Yeah. In this case, I am going in circles, but what if I told you that someone else had already calculated the ratio of the sides of all angles between 0 and 90? And what if I told you that you can use those ratios to calculate the length of the conduit travel when you only know the angle of the bend and the length of the offset?" Tom showed them a table of cosecant values at various angles.

ANGLE	COSECANT	ANGLE	COSECANT	ANGLE	COSECANT	ANGLE	COSECANT
1°	57.3065	24°	2.4585	47°	1.3673	70°	1.0641
2°	28.6532	25°	2.3661	48°	1.3456	71°	1.0576
3°	19.1058	26°	2.2811	49°	1.3250	72°	1.0514
4°	14.3348	27°	2.2026	50°	1.3054	73°	1.0456
5°	11.4731	28°	2.1300	51°	1.2867	74°	1.0402
6°	9.5666	29°	2.0626	52°	1.2690	75°	1.0352
7°	8.2054	30°	2.0000	53°	1.2521	76°	1.0306
8°	7.1854	31°	1.9415	54°	1.2360	77°	1.0263
9°	6.3926	32°	1.8870	55°	1.2207	78°	1.0223
10°	5.7587	33°	1.8360	56°	1.2062	79°	1.0187
11°	5.2408	34°	1.7883	57°	1.1923	80°	1.0154
12°	4.8097	35°	1.7434	58°	1.1791	81°	1.0124
13°	4.4454	36°	1.7012	59°	1.1666	82°	1.0098
14°	4.1335	37°	1.6616	60°	1.1547	83°	1.0075
15°	3.8636	38°	1.6242	61°	1.1433	84°	1.0055
16°	3.5915	39°	1.5890	62°	1.1325	85°	1.0038
17°	3.4203	40°	1.5557	63°	1.1223	86°	1.0024
18°	3.2360	41°	1.5242	64°	1.1126	87°	1.0013
19°	3.0715	42°	1.4944	65°	1.1033	88°	1.0006
20°	2.9238	43°	1.4662	66°	1.0946	89°	1.0001
21°	2.7904	44°	1.4395	67°	1.0863	90°	1.0000
22°	2.6694	45°	1.4142	68°	1.0785		
23°	2.5593	46°	1.4395	69°	1.0711		

Key To Understanding

The angles of any right triangle will determine the lengths of its three sides. When you know the size of one angle and the length of one side of a right triangle, you can calculate the lengths of the other sides using trigonometric functions. These functions are ratios of the lengths of the sides of the triangle. For more information about trigonometric functions, see Chapter 14.

Math Speak

Trigonometry is the study of triangles. Trigonometric functions define relationships between the lengths of the sides of a right triangle and their angles.

The students sat and thought. The shop was very quiet for a minute. Then Al broke the silence, pointing to the picture shown below. "Are you telling me that all I need to know is the angle of the bend of the conduit and the offset depth to figure out the length of the travel?"

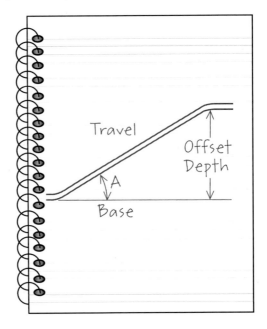

"Yep," Tom said. "It's called trigonometry. The ratios are called functions. And the offset constant is called the cosecant. When you know the value of angle a and the length of the offset, you look up the cosecant in this table and then multiply it by the offset to get the travel."

Chapter 10 Review Problems

1. A circle has a radius of 8 inches. What is the diameter? What is the circumference?

    r = 8"
    d = 2r = 16"
    c = πd = 3.14 × 16" = 50.24"

2. A circle has a circumference of 201 inches. What is the radius?

    c = πd
    c = 201" = 3.14 × d
    201" ÷ 3.14 = (3.14 × d) ÷ 3.14
    64" = d
    r = (½)d
    r = 32"

3. The sum of angles b1 and b2 equal 82 degrees. Angle a1 equals 48 degrees. Calculate the size of the angles in the following figure.

    Sum of b1 and b2 = 82°

    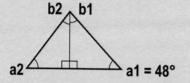

    a1 = 48°

    ∠a1 = 48°
    ∠b1 = 42° (90° − 48°)
    ∠b2 = 40° (82° − 42°)
    ∠a2 = 50° (90° − 40°)

4. The ratio of x to y is 2:1.

    When x is 4, y is **2**.

    When x is 16, y is **8**.

    When x is 1, y is **½**.

Lots of students have trouble understanding how to use the cosecant function. Let's try a couple of problems. Use the following figure to complete Questions 5 through 9.

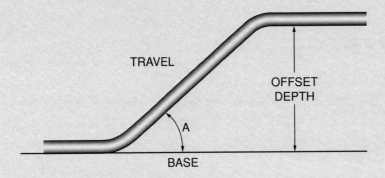

5. Find the length of travel for a piece of conduit that has a 29-degree bend and an offset of 10 inches.

    First, use the table to look up the cosecant for 29 degrees. It's **2.0626**.

    Next, multiply the offset by the cosecant to find the travel. 10 inches times **2.0626** equals **20.626** inches.

6. Find the length of travel for a piece of conduit with a 25-degree bend and an offset of 20 inches.

    Cosecant = 2.3661
    2.3661 × 20 = 47.322 inches

7. Find the length of travel for a piece of conduit with a 45-degree bend and an offset of 15 inches.

    Cosecant = 1.4142
    1.4142 × 15 = 21.213 inches

8. Find the length of travel for a piece of conduit with a 60-degree bend and an offset of 5 inches.

    Cosecant = 1.1547
    1.1547 × 5 = 5.7735 inches

9. Find the length of travel for a piece of conduit with an 11-degree bend and an offset of 10 inches.

    Cosecant = 5.2408
    5.2408 × 10 = 52.408 inches

Careers in Construction—Electrician

Electricians design, install, and repair wiring and other electrical components (such as breaker boxes, switches, and light fixtures) in homes and commercial buildings. Electricians usually need to be licensed, and they need to know a lot of electrical theory, which includes math. They also need to know local and national electrical codes. It takes a long time—usually around four years—and a lot of knowledge to become a good electrician. Since electricians work around electricity, the work can be dangerous. Electricians need to be very safety conscious and willing to follow very exacting rules. A bad electrical job can kill someone! If you want to know more about electrical work, check at your local community college or look on the National Center for Construction Education and Research website (www.NCCER.org).

Did You Know

Ponnequin Wind Farm generates electrical power from the wind. The site is located on the plains of eastern Colorado just south of the Wyoming border. It consists of 44 wind turbines and can generate 30 megawatts of electricity.

Each wind turbine costs about $1 million to build and is capable of generating 700 kilowatts of electricity. The turbines begin operating with wind speeds as low as 7 mph and shut themselves down at speeds over 55 mph to prevent overspeed damage. Each turbine weighs nearly 100 tons and stands 181 feet tall. The turbine blades are attached on top of the turbine and have a diameter of 159 feet.

NOTES

Chapter 11
First I'm Hot, Then I'm Cold

There's a lot of math to HVAC systems, but most of it is used to design the system rather than install or maintain it. You can demonstrate how the size of HVAC ducts impact heating and cooling with an electric fan. Tape a large plastic bag around the perimeter of the fan and increase and decrease the size of the aperture to change the force of the air from the fan. Demonstrate how the fan needs a complete path of circulation by using a stiff piece of cardboard to block the airflow at the back of the fan. Be sure to set up the demonstration with the fan off.

Demonstrate the concept of Btus by using several kitchen matches to heat water. One Btu heats one pound of water one degree Fahrenheit. It is hard for anyone to believe that 1 Btu, which is approximately the amount of heat generated when 1 wooden kitchen match is completely burned, will raise the water a degree, so you can demonstrate it. If you have access to a chemistry lab, you can perform a very controlled demonstration, but you can still demonstrate this with common household materials. Before you begin, explain to the students that this demonstration can't capture all of the heat from the kitchen match, so the water won't increase 1° per match.

1. Place 1 pound of water, which is slightly more than 1 cup (1 gallon of water equals 8.33 lbs), but you should measure the weight on a scale—seeing is believing, in a flameproof container (an empty food can will do—just be careful not to burn yourself with the flame.)

2. Use a candy thermometer or other thermometer to measure the temperature of the water. Remove the thermometer.

3. Place five or more wooden kitchen matches (the long ones) in channel locks, vise grips, or other suitable device. The number of matches you use will depend on the sensitivity of your thermometer, but use more than one, since some of the heat is lost to the air.

4. Light the matches and hold them under the container of water until they burn out. (Hold them as close to the container as you can so you don't lose heat.)

5. Use the thermometer to gently stir the water and then measure the temperature again. The temperature will have increased.

NOTES

Chapter 11

First I'm Hot, Then I'm Cold

"Attention, please," Al began. "You'll be very pleased to hear that there is very little math in heating and air conditioning installation."

A series of cheers went up from the students sitting on the floor of what would be the Browns' living room.

"Silence," Al said. "Unfortunately, there is a lot of math involved in the design of heating and air conditioning systems."

The class responded with a chorus of groans.

"But, for once, and only once, Mr. Whyte says we don't need to learn all of it."

"Al," Mr. Whyte sighed. "You can take a five-minute story and turn it into an hour. Move it along."

"You bet, Mr. Whyte. We're talking about heating, ventilation, and air conditioning—it's abbreviated HVAC. Here are some brochures about a bunch of different systems. Notice that some of these brochures talk about both heating and air conditioning, and some talk about just heating. That's 'cause in some places, not every one has air conditioning. In Florida, we're more worried about air conditioning than heating, but up north, they're more worried about heating.

Key To Understanding

Volume is the amount of space a solid figure—like a slab—occupies. The solid is sometimes called an object, and it has three dimensions: length, width, and depth (sometimes depth is called height). Volume is calculated by multiplying its length, width, and depth.

Rectangular object or solid
Volume = L × W × D

Cube
Volume = S^3

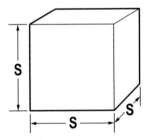

"Sometimes when you're installing a heating and air conditioning unit, you need to pour a concrete slab to set it on. We all know how to figure out volume of a slab—it's length times width times depth."

"In residential construction," Al continued, "you'll probably use a pre-formed slab for the heating and air conditioning unit, so you don't need to figure out the volume."

"Hey, this is great. I caught everything so far," Travis said.

Mr. Whyte laughed. "Hit 'em with your smarts, Al."

"Pay close attention. Anything that burns—coal, gas, oil, and wood—can produce heat, and the amount of heat produced is called heat content. Heat content is measured in Btus."

"I've heard of Btus, Al, but what are they?" Phil asked.

"Btu is the abbreviation for British thermal unit," Al continued. "In case you can't figure it out for yourself, the name came from England. A Btu is the amount of heat it takes to raise 1 pound of water 1 degree Fahrenheit. This handout shows you the number of Btus in some things."

Approximate Btu Values of Selected Energy Sources

1 wooden kitchen match = 1 Btu
1 gallon of gasoline = 125,000 Btus
1 gallon of heating oil = 139,000 Btus
1 cubic foot of natural gas = 1,021 Btus
1 gallon of propane = 91,000 Btus
1 pound of coal = 10,000 Btus
1 kilowatt-hour of electricity = 3,412 Btus
1 pound of wood = 6,200 Btus

"When you talk about a heating system, you are not really adding heat to water. You're adding heat to the air in the house, and when you talk about air conditioning, you're talking about removing heat from the house—still in Btus. So any unit that has a high Btu rating can heat or cool better than a unit with a low Btu rating."

 **Did You Know**

In the United States, electric power is generated in many ways, but almost 50% of the electric power in our country is generated by burning coal. A well-designed modern electric power plant can make one kilowatt-hour of electricity from about a pound of coal.

What does that mean to you? Well, almost a pound of coal is needed to make one kilowatt-hour of electricity, and one kilowatt-hour of electricity is enough to run a 60-watt light bulb for about 15½ hours.

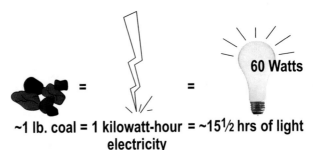

~1 lb. coal = 1 kilowatt-hour = ~15½ hrs of light
electricity

The chart below shows some of the other ways used to make electricity. (The chart is from the Department of Energy, which is a department of the U.S. government.)

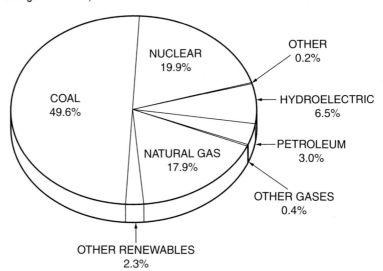

According to the U.S. Department of Energy website, it takes more than 2,000 kilowatt-hours of electricity a year to run a central air conditioning unit for the average house. (Central air conditioning means one big unit cools the whole house.) That's a lot of coal!

You can get answers to all sorts of energy questions at the Department of Energy websites (www.energy.gov and www.eere.energy.gov).

If you have specific questions about energy or any other science or math problem, you can submit a question to Ask a Scientist at Argonne National Laboratory in Illinois using their Division of Educational Programs (DEP) website (www.dep.anl.gov). Argonne National Laboratory is part of the U.S. Department of Energy and does a lot of research about energy and other sciences.

Key To Understanding

Lots of people don't understand about Btus. A Btu is just a name that we give to the amount of heat it takes to raise (or lower) the temperature of 1 pound of water 1 degree Fahrenheit. A Btu is just a unit of measure. It's something that we all agree to call that unit of energy.

In the United States, we commonly measure temperature in degrees Fahrenheit. The Fahrenheit temperature scale was named after the German scientist, Daniel Gabriel Fahrenheit, who developed it. In this system, the freezing point for water is 32 degrees and the boiling point is 212 degrees. There are lots of stories about how Mr. Fahrenheit arrived at these numbers, but the point is that they are numbers that he selected and that we all agree to use—in the United States, that is!

In other countries, temperature is measured in degrees Celsius. The Celsius scale was named after the man who designed it. His name was Anders Celsius. Mr. Celsius defined 0 degrees as the freezing point of water and 100 degrees as the boiling point. We all agree to call the freezing point of water 0 degrees Celsius.

If you need to convert temperatures between scales, use the following formulas:

Temperature in Celsius = (5/9) × (Temperature in Fahrenheit – 32)

Temperature in Fahrenheit = (9/5) × (Temperature in Celsius + 32)

"So the higher the Btu rating of a unit, the better. Right, Al?" Tom asked.

"No," Al said. "Using a unit that's too powerful is just as bad as using a unit that's too weak. In this part of Florida, we don't need powerful heating units because it doesn't get that cold here, and when it does get cold, it doesn't last, so a powerful heater would hardly run. But in Maine, it gets really cold for a long time. A small heating unit would need to run a lot to heat a house, and that's not good. There are lots of reasons. First, a big unit costs a lot. Second, every unit has a best run time—that means that a unit needs to run for a minimum period of time to be at its most efficient. If you use a unit that's too big, the run time is too short, but with a unit that's too small, the run time is too long. Third, you need to remember that when an undersized unit runs for a long time, it is using electricity. Finally, there are all sorts of other things to consider—like an air conditioner is supposed to take humidity out of the air as the air is being cooled. If the unit runs for too short of a time, it can't take as much humidity out of the air."

"It's really important to talk to an HVAC professional when you select units for new construction or replace units in an older home," Mr. Whyte pointed out.

Al held up a picture and continued. "It doesn't take a genius to figure out that when you want to heat a house, you need to somehow warm the air. In the Browns' house, we're using a forced air system. That means we'll pull air out of the house, heat it in the heating unit, and then push it back into the house."

"Let's think about this for a minute," Mr. Whyte said. "Let's say that we selected the best heater for the house. What's the one thing that can make or break whether the unit will heat the house comfortably?"

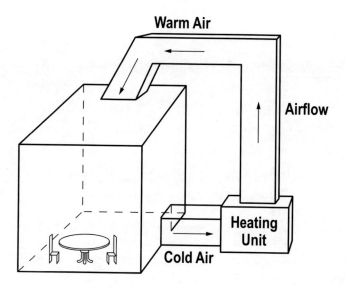

The class thought about this. Some flipped through the brochures that Al had handed out earlier.

"Come on, guys," Mr. Whyte said. "Think about the airflow. What affects the airflow in an HVAC system?"

"Oh," Olivia said. "The ductwork."

Al smiled. "Right-o, Olivia. The size of the ducts affects the volume of air that can circulate through the HVAC system." Al held up the picture again. "Even though you can't see it, a room is filled with air, so it has volume."

"And the ducts are three dimensional, so they have volume." Al went on. "Using undersized ducts will decrease the airflow through the heater unit and decrease the efficiency of the unit. Using undersized ducts is as bad as using an undersized unit.

"The good news is that we don't need to figure out all of this stuff for HVAC systems. Engineers figure it out. Our job is to follow the HVAC specifications, so the system can work at peak efficiency," Al finished.

Chapter 11 Review Problems

Mrs. Brown found a central air conditioning unit on sale. The unit is smaller than what is needed to cool the house. You are holding the following conversation with her and are trying to convince her that buying the unit is not a good idea. Write short answers in your own words that answer Mrs. Brown's questions. Hint: Keep it simple.

Mrs. Brown: I was over at Central Air HQ and found a heating and air conditioning unit on sale. I'm thinking of buying it. What do you think?

You: It's okay as long as it meets the specifications in the HVAC plans.

Mrs. Brown: It doesn't exactly. The air conditioner that's on sale is rated at 36,000 Btus an hour and the specification says it should be 42,000. What's a Btu?

You: **It's a unit of measure for heating and cooling. The higher the Btu rating, the higher the unit capacity**.

Mrs. Brown: So if I buy this unit that's on sale, what will happen when it gets hot outside?

You: **The air conditioner will need to run more often and for a lot longer than a larger unit to cool the house.**

Mrs. Brown: How will that affect my electrical bill?

You: **During the summer, your bill will be higher because the unit is out of its best operating range.**

Mrs. Brown: When you put it that way, I can see that buying the smaller unit isn't such a good deal.

Instructor Note: Answers will vary.

Careers in Construction—HVAC Industry

Our physical comfort is important to us, so people go to great lengths to stay warm and keep cool. The HVAC industry has a wide variety of careers. Some of these jobs require a lot of math knowledge, but some require very little. HVAC engineers need to know a lot of math and physics to design an efficient and effective HVAC system, while an installer needs to have mostly mechanical skills. HVAC technicians need to have skill levels in between.

You can break into the HVAC industry in two ways. A company can hire you and give you on-the-job training (OJT) or you can enroll in a formal education program in which you attend classes and get hands-on training. In either case, it's important for those in the HVAC industry to be responsible and reliable because some of the chemicals that are used in air conditioning systems are damaging to the environment. These chemicals are called chlorofluorocarbons (CFCs) and hydrochlorofluorocarbons (HCFCs). These chemicals were once widely used, but were found to be so dangerous that the U.S. government has restricted their use. They are so dangerous that anyone who releases them into the atmosphere can be fined or go to prison!

History—HVAC

One of the neighbors stopped by to see the Browns' house. Her name was Edith Wilson and everyone called her Miss Edith (in the southeastern part of the United States, men and women are often addressed with the titles Miss or Mister before their first name). She was in her 80s and was born in Miami but moved to this area when she was a little girl. She told the students about living in rural Florida when there was no electricity—that means no air conditioning! "Back then, we spent our summers outside," she said. "Sometimes we even cooked outside. At night we would all sleep on the porch, because it was so hot in the house."

Before the widespread use of air conditioning, houses in Florida were designed differently. They were often built off the ground with a crawl space beneath them. This helped to improve air circulation and to keep bugs out of the house. Since hot air is lighter than cool air, it rises to the highest point in the house, so houses were built with high ceilings and tall double-hung windows (that means the top and bottom windows slide up and down). In the summer, people would open the top of the window to allow hot air to escape.

People who were very fortunate had screened in porches, called Florida rooms. These porches allowed the family to sit outside and not be bothered with bugs. Back then, people often sat out on their porches at night and read, played games, or listened to the radio—on some hot summer days, they even made homemade ice cream!

In the winter, people often heated with woodstoves and fireplaces. There is a saying that wood warms you twice—once when you chop it and once when you burn it. It takes a lot of time and energy to heat with wood. First, you need to cut or buy the wood. Next, it needs to be stacked, and then when it's time to burn it, you need to haul it into the house. Once the fire is burning, it needs to be fed to keep it going. At night, the fire needs to be banked (covered with embers) to keep it going until morning.

Chapter 12
Inside and Out

Often students have no trouble performing calculations, but they have trouble setting up the problem. Just as Mrs. Brown thought that the surface area of a room's floor would be the same as the walls, students frequently have trouble applying math skills to real life. You can help them by challenging them with everyday problems. Here are some examples:

Look at a pitched roof. What are some formulas that can be applied to the roof? Examples include the surface area of the sheathing, volume of the attic space, slope, pitch, and the Pythagorean theorem. Next, ask the students when they need to use the formulas.

Look at stormwater drainage in a parking lot. Why do some parking lots flood in a heavy rain? Runoff is measured in volume; more rain produces more volume, but the drain size stays the same. Pose this question: if the parking lot was enlarged, but the drainage system was not, what would be the outcome?

NOTES

Chapter 12

Inside and Out

"I need to get paint for the walls and I know what colors I want, but I don't know how many gallons of paint to buy. The clerk at the store said one can of paint covers about 400 square feet," Mrs. Brown told Travis and Olivia. "I want to paint the living room and dining room the same color. How much should I buy?"

Travis and Olivia looked at each other. The look said, "Oh, no. Not again." Mrs. Brown kept changing her mind. Yesterday she had Travis use strings to outline two rectangular flowerbeds in the back yard, and then she changed her mind and wanted four square ones.

Yesterday she wanted to paint the dining room green and the living room blue. Now she wanted to paint the dining room and living room the same color.

Olivia picked up her notebook and flipped to the diagram she had drawn yesterday. "Remember how yesterday we said that in the dining room each wall was really a rectangle? The ceiling is 9 feet high so the dining room walls make two rectangles that are 15 feet by 9 feet and two that are 12 feet by 9 feet. You can ignore the doors and windows in an area this small—they affect the amount of paint you need, but not by much."

Math Speak

Surface Area

Area is how much space there is on a surface. It's measured in square units like inches, feet, and yards. Surface area for a square or a rectangle is calculated as follows:

Length × Width = Surface Area

The equation can be written with abbreviations as follows:

L × W = SA

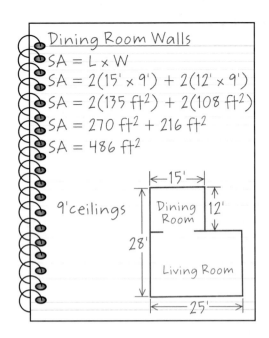

"You can calculate the surface area using the perimeter of the room and then multiply it by the height of the walls," Travis said. "The living room is 25 feet by 16 feet, so the perimeter is 82 feet. The area of the living room walls is 738 square feet—82 feet times 9 feet. And the total wall area of both rooms is 1,224 square feet."

Key To Understanding

A yard is 3 feet long. A square yard is 3 feet by 3 feet, so a square yard is equal to 9 square feet (3 feet times 3 feet equals 9 square feet). To convert square feet to square yards, divide by 9.

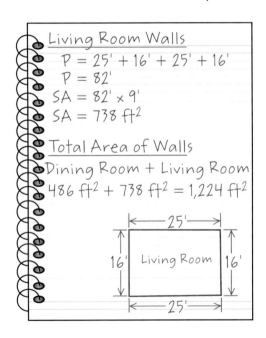

"Got it. 1,224 square feet. Is that number good for the carpet, too?" Mrs. Brown asked.

Travis and Olivia looked at Mrs. Brown. Olivia said, "Excuse me?"

"The carpet. I want to order carpet for the floor, too. It comes in square yards so can I tell the clerk at the store to give me whatever 1,224 square feet is in square yards?"

"Sure. If you're going to carpet the walls!" Travis said. Olivia elbowed Travis in the ribs.

"Well, Mrs. Brown," Olivia began. "You can convert square feet to square yards, but this number is the area for the dining room and living room walls, not the floor. Since the carpet goes on the floor, we need to calculate the area for the floor space."

Mind Games

Did you notice that Travis used 16 feet for one of the living room dimensions? Olivia's diagram shows the wall on the left side of the figure is 28 feet, but that includes the dining room wall. To calculate the length of the two living room walls, Travis subtracted the length of the dining room wall, which is 12 feet, and got 16 feet for the length of the living room wall (28 feet minus 12 feet equals 16 feet).

Practice Problems 12-1

1. Calculate the number of square feet of carpeting Mrs. Brown needs for the living room and dining room.

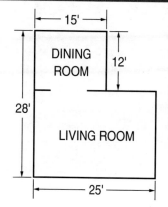

Dining room = 15' × 12' = 180 square feet
Living room = 16' × 25' = 400 square feet
Total = 400 + 180 = 580 square feet

2. Convert the number of square feet of carpeting into square yards.

580 square feet ÷ 9 = 64⅔ square yards

3. Mrs. Brown changed her mind. She decided to tile the dining room and carpet the living room. How much tile does she need in square feet and in square yards? How much carpeting (in square feet and in square yards)?

Tile: 180 square feet = 180 ÷ 9 = 20 square yards
Carpet: 400 square feet ÷ 9 = 44⅔ square yards

Mrs. Brown smiled at Olivia. "Thanks so much. I was good at math in school, but I never learned how to apply it to real life."

"No kidding," Travis said under his breath. Olivia stepped in front of Travis. Over the months that they had been working together, Olivia discovered that Travis had no patience when he had to deal with Mrs. Brown. He thought she asked too many questions.

"Hey," someone shouted from the garage. "Where is everyone?"

Travis, Olivia, and Mrs. Brown turned towards the voice as Phil came through the door, carrying a notebook and large framing square. He stopped when he saw Mrs. Brown.

"Sorry to interrupt, Mrs. B," Phil said. "I didn't know you were here."

"We were just finishing up. I'm glad to see you, Phil," Mrs. Brown said. "Mr. Whyte told me that you might be able to use the roof sheathing that was leftover from the house to roof the picnic pavilion."

"Yes, ma'am," Phil said. "It depends on what pitch you want for the roof."

"What's pitch," Mrs. Brown asked. "Is that the same as slope?"

"Almost but not exactly," Phil responded. "I'll show you using some handouts Mr. Whyte gave us in class. Slope is the ratio of rise to run. It's always given for a 12-inch run, so an 8-inch rise over 12 inches has an 8-12 slope."

He pointed to the diagram. "Pitch has to do with the entire span of the roof and is always a fraction. A gable roof with an 8-12 slope has a pitch of 8 over 24, which reduces to $\frac{1}{3}$."

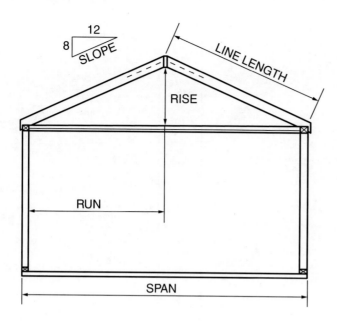

"Wait a minute, Phil. Where did you get 24?"

"It's twice the run," Phil said, pointing to the diagram. "This roof is very simple. It has two slopes that meet in the center."

"It's so hard to see it," Mrs. Brown sighed.

"No problem, Mrs. Brown," Phil said. "This is a framing square. It's marked with inches on both sides of the angle." Phil knelt on the bare concrete floor and laid the framing square in front of him. Travis and Olivia leaned over to see what Phil was showing Mrs. Brown. Phil used his tape measure to show Mrs. Brown the difference between pitch and slope. "In roofing, the run is always 12 inches and it's called a unit of run. A 12-inch run with a rise of 24 inches has a slope of 24-12, but the pitch is 1, because the pitch is the rise over the span and the span is 24—12 inches for each side of the roof."

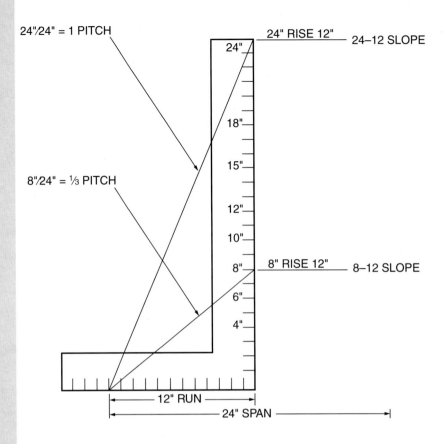

"That would be a really steep roof," Mrs. Brown said.

"It sure would." Phil moved his tape measure to show Mrs. Brown several other roof slopes and pitches. "When you talk about 12 inches of run, that's called a unit or run. To figure out the total rise, you need to multiply the rise by the total run of the roof."

"I don't get it. How does that tell you the total rise?" Mrs. Brown asked. Travis let out a sigh.

Phil shot Travis a warning glance and then said to Mrs. Brown, "The pavilion is 10 feet wide—that's the total span. The run is half of that, so it's 5 feet. Let's say we'll use a 4-12 slope." Phil laid his tape measure across the framing square to show a 4-12 slope. "If I know I have a rise of 4 inches for every 12 inches—or one foot—of run, then across a 5-foot run I must have 20 inches of rise.

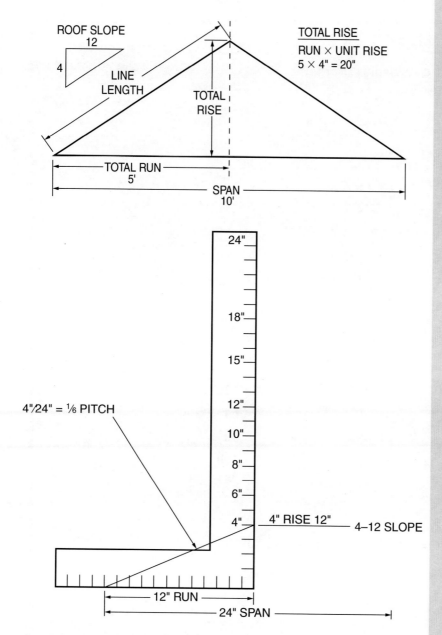

The pitch is ⅙ because 20 divided by 120 inches is ⅙."

Mrs. Brown frowned. "Where did you get the 120 inches from?"

"I converted the 10-foot span to 120 inches," Phil replied.

"Of course," Mrs. Brown smiled. "That makes sense."

Practice Problems 12-2

Complete the following problems.

1. Find the pitch for the following roofs:

Total Rise	Span	Pitch
8'	24'	8/24 = 1/3
6'	36'	6/36 = 1/6
9'	24'	9/24 = 3/8
12'	36'	12/36 = 1/3
8'	32'	8/32 = 1/4

2. Find the total rise for the following roofs:

Pitch	Span	Rise per Foot of Run	Total Rise
1/2	24'	12"	144" = 12'
1/8	32'	4"	64" = 5'-4"
1/3	24'	8"	96" = 8'
5/12	30'	10"	150" = 12'-6"
3/4	24'	18"	216" = 18'

3. Find the rise per foot of run for the following roofs:

Span	Total Rise	Rise per Foot of Run
18'	6'	6/9 = 2/3 = 8/12
24'	8'	8/12 = 2/3 = 8/12
28'	7'	7/14 = 1/2 = 6/12
16'	4'	4/8 = 1/2 = 6/12
32'	12'	12/16 = 3/4 = 9/12

"When I know the total run and the total rise, I can use the Pythagorean theorem to determine the length of the roof line and that tells me how long my roof sheath needs to be," Phil went on.

"That Pythagorean theorem gets a workout in construction," Mrs. Brown said. "It seems like you're always using it. Do you have enough sheathing to finish the roof? Maybe I should order more. I'm stopping at the lumber yard on my way to the paint store."

"Let's see," Phil said. "We are using an 8-12 slope on the pavilion roof, so 8 inches times 5 is 40 inches—that's 3 feet 4 inches of rise." He showed her a drawing of the pavilion.

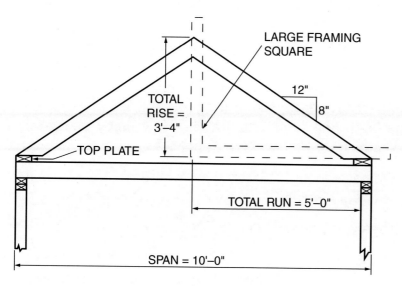

He took out his notebook and began making calculations. "Using the Pythagorean theorem, the roof line length comes out to just a bit over 6 feet."

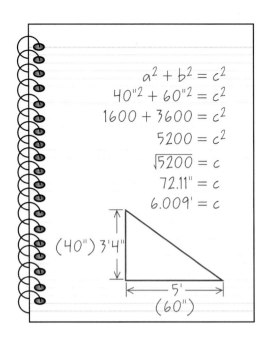

"The roof sheathing is 4 feet wide by 8 feet long. We'll cut the sheathing to 6 feet so length is no problem. The covered part of the pavilion is 12 feet wide, so we need 3 pieces for each side of the roof. That's 6 pieces altogether." Phil turned to Travis. "How many panels of sheathing do we have?"

"We have enough," Travis said. "There are 4 pieces that are 6 feet long and 6 pieces that are 8 feet long."

"Then we're all set," Phil told Mrs. Brown.

"Good, I won't order more roof sheathing." Mrs. Brown turned to Travis. "Travis, I love the flower beds in the back yard, but I've decided to go back to the rectangular one that you laid out first. And I want circular ones, too. Here's a diagram. Can you figure out how much **mulch** I'll need for a 6-inch layer? I'll call you when I get to the lumber yard—can you believe it! A lumber yard sells mulch, too!"

Mrs. Brown waved and walked away. Travis stared after her and then said, "Nope, pretty unbelievable."

 **Did You Know** You can apply the same mathematical functions you used in roofing to stairs. It's a little more complicated because each stair is the unit—it's called the tread. The tread rise and run both vary in length, depending on space limitations and local building codes, with codes taking precedent. Generally, an 11-inch run and 7-inch rise is considered the most comfortable for a person to walk up and down. In stairs, the total rise is the distance between the finished surfaces of the upper and lower floors. The total run of a stairway is determined by multiplying the tread run by the number of stairs.

TOTAL RUN = # TREADS × TREAD RUN
TOTAL RUN = 13 × 9$^{15}/_{16}$"
TOTAL RUN = 129$^{3}/_{16}$"

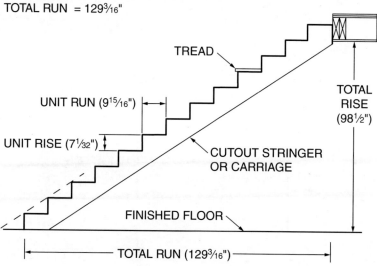

 **Math Speak**

When a circle has volume, it is called a cylinder. You calculate the area of a circle using the formula:

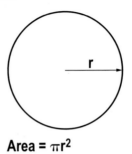

Area = πr^2

You calculate the volume of a cylinder using:

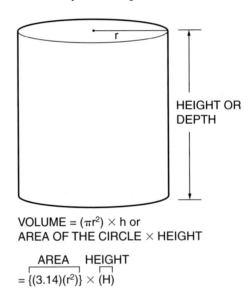

VOLUME = $(\pi r^2) \times$ h or
AREA OF THE CIRCLE × HEIGHT

$$= \underbrace{\{(3.14)(r^2)\}}_{\text{AREA}} \times \underbrace{(H)}_{\text{HEIGHT}}$$

Volume is measured in cubic units—like cubic inches, feet, and yards.

Chapter 12 Review Problems

1. Mrs. Brown forgot to have Olivia calculate the number of square feet of tile she needs for the kitchen. The kitchen is 12 feet by 10 feet. How much tile should Mrs. Brown buy in square feet and in square yards?

   12' × 10' = 120 square feet
   120 square feet ÷ 9 = 13⅓ square yards

2. Mrs. Brown thought it would be a good idea to make the pavilion roofline 8 feet long so the students wouldn't need to cut the sheathing. Calculate the total rise and the roof slope and pitch. The total run is 5 feet so the span is 10 feet. Hint: Use the Pythagorean theorem to calculate the total rise, and then divide the total rise by the total run to get the unit rise.

   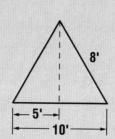

   $a^2 + b^2 = c^2$
   $a^2 + 5^2 = 8^2$
   $a^2 + 25 = 64$
   $a^2 + 25 - 25 = 64 - 25$
   $a^2 = 39$
   $a = \sqrt{39}$
   $a = 6.25$ ft
   Rise = 6¼ ft
   Rise per foot = 6¼ ÷ 5
   Rise per foot = $\frac{25}{4} \div 5 = \frac{5}{4} = 1¼$ ft

3. Here's the diagram Mrs. Brown drew of the garden. Calculate the amount of mulch needed for each bed when the depth of mulch is 6 inches.

   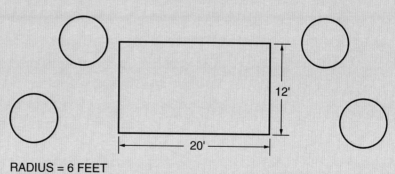

   RADIUS = 6 FEET

   Rectangle = 20' × 12' = 240 square feet

   Volume = 240 square feet × 6/12 = 120 square feet = 120 ÷ 27 = 4¹²⁄₂₇ cubic yards

   Circle = 3.14 × 6² = 3.14 × 36 = 113.04 square feet

   Volume = 113.04 square feet × 6/12 = 56.52 cubic feet = 56.52 ÷ 27 = 2.09 cubic yards

 **Terms**

Mulch: A layer of material, such as sawdust, bark chips, pine straw, or paper, that is spread on the ground in a garden. Mulch helps keep the ground cool, controls weeds, and conserves water. It also protects the ground from erosion. Most mulch is made from vegetative matter, so it adds nutrients into the soil as it breaks down.

 Careers in Construction— Interior Design and Landscape Design

Interior designers are able to match furnishings, floor and window coverings, and colors to rooms. Homeowners tell the designer what type of lifestyle they have and what sort of things they like, and the designer decorates the house toward these preferences. Interior designers are good at matching colors and textures, and they need to be good listeners so they can give the homeowner what they want. Some people go to college to become an interior designer, but others learn by doing it.

A landscape designer selects plants and trees for the homeowner. Landscape designers need to know a lot about plants—where they grow best, how big they grow, and how much water and sun they need. Some landscape designers go to college. These people usually design big projects like parks and public gardens. Other landscape designers attend short classes, which are often offered by the local Extension Service, or they learn by self-study. If you want to be a landscape designer, you should like working outside.

Extension services are informal education programs that are offered to the public, usually for a small fee. Sometimes a government agency will have extension services but often it is the local colleges and universities.

Chapter 13
The Bottom Line

Jorge loves accounting, but most people are only interested in their paycheck. All expenses for a job (any job) need to be taken out of revenue; when expenses exceed revenue, the business can go bankrupt. When expenses equal revenue, there is no profit. Profit is often invested back into the company to help it grow. Part or all of the profits can also be dispersed among investors in the company.

Some expenses are easy to identify—salaries, materials, equipment, etc. Others are not so easy to see—insurance, electricity, rent, office supplies, etc. Depending on the age of your students, you may be able to relate hidden expenses to the price of gasoline. When the price of fuel is high, it affects everyone at the pump. Ask your students to think about how the price of gasoline affects the price of a loaf of bread or a gallon of milk.

Go further. Explain how our gasoline is made from oil. According to the US Department of Energy website, oil is used to generate 40% of US energy needs, and it generates 99% of our transportation needs. Most of the gasoline that we use is refined in the US, and about 45% of it comes from refineries located in the Gulf States. That's why when hurricanes hit the Gulf States, gasoline supplies are decreased and sometimes the prices increase. Even if the hurricanes don't do damage, refineries are usually shut down as a precautionary measure.

In 2004, the US imported approximately 60% of the oil it consumed. Ask your students what this means to our country and what we can do about it. Obvious answers are conserve energy (use less gasoline and electricity), use alternative energy (great in theory, difficult in practicality), and allow further exploration in the US. Most people believe that the US gets most of its imported crude oil from Saudi Arabia. This is not true. Canada is our top supplier, followed by Mexico (www.eia.doc.gov).

NOTES

Chapter 13

The Bottom Line

Jorge sat on a tall stool in front of the workbench in Mr. Brown's shop. There were neat piles of colored **invoices** stacked on the workbench in front of him. He was entering figures on the invoices into the accounting program on a notebook computer. As he entered the figures from an invoice, he placed a checkmark next to the figure. When he entered all of the figures on the invoice into the computer, he scribbled the word Completed at the bottom of the invoice and then wrote his name and the date.

Mr. Whyte, Mr. Conrad, Tom, and Al walked into Mr. Brown's shop. They all looked expectantly at Jorge. Mr. Whyte asked, "How's our bottom line looking, Jorge?"

Jorge threw his pencil onto the workbench and leaned back, stretching his arms over his head. "I still have to estimate payroll until the end of the project, but I have all other costs in and it looks good. Mr. Brown is giving us a $1,000 bonus because Tom saved him so much on the electrical."

Tom grinned widely. Mr. Conrad threw his arm around Tom's shoulders and said, "That's my boy!"

Mr. Whyte looked at Mr. Conrad and said, "Chip off the old block, eh?"

"Nah, Tom got his looks from me," Mr. Conrad said, ruffling Tom's hair. "But he was lucky enough to get his mother's brains."

They all laughed. Al turned to Jorge and said, "How do you figure out if we made a profit? We have deliveries coming in here and stuff being returned so fast I can't keep up."

"It's a lot of paperwork, but once you get a system going, it's easy to track," Jorge said. "When Mr. Whyte orders materials, he sends me a copy of the order form—that's called a purchase order. Later, when the supplier delivers the goods, whoever takes delivery checks the purchase order against the invoice—sometimes the invoice is called the packing slip. I get the invoice and double check it to be sure that the price the supplier charged us is right. Then I enter the numbers into the accounting program, so I can track costs."

"You need to enter the cost for every nail and wire and can of paint we buy?" Al asked.

"Not just that," Jorge continued. "On payday, I enter everyone's salary and the payroll taxes that the school pays the government, too. Those are part of direct costs."

"What are direct costs?" Tom asked.

"Direct costs are things that are used on the job. Plumbing supplies, lumber, nails, construction workers' salaries, the insurance the school needs to pay for each student working on the site—those are direct costs. Indirect costs are things we need but aren't used to build the house. Like the school secretary checks our time cards and sends them to the accounting department so we can get paid. The accountant enters our hours into the payroll program and writes checks for us, but they don't come down to the site and work on the house. Part of their salaries are indirect costs."

"Gotcha," Al said. "So where's the profit, man?"

"After I get all of the direct and indirect costs together, I subtract them from the amount the Browns paid for the house. Anything left over is profit."

Key To Understanding

It's really important for everyone on a job site to take accounting seriously. The flow for purchase orders on a typical job site is shown below. The person accepting a delivery must be certain to check that the order is complete and the goods are not damaged. Never sign for an order without checking it first.

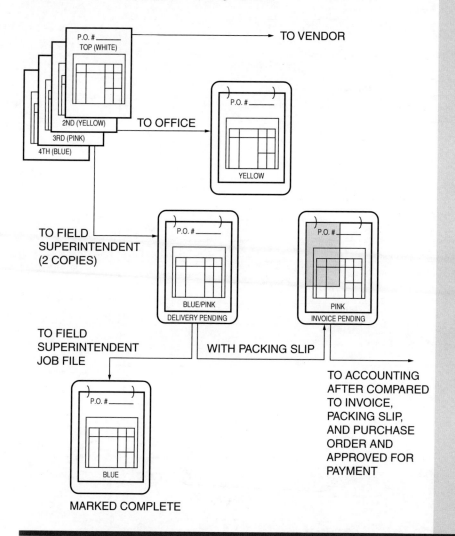

Chapter 13 Review Problems

1. Jorge received the following invoice from the local home improvement store. Calculate the costs for items 1, 2, and 6.

		INVOICE		
Homes For U	Date: 3/17			
	Invoice Number: 171422			
Bill To: Loften High Industrial Arts	Ship To: Brown 602 NE Limerock Road Brooker, FL			
Item No.	Quantity	Description	Price	Amount
1	1500 ft²	Carpet per sq ft	2 | 62	**3930 | 00**
2	500 ft²	Ceramic tile per sq ft	2 | 79	**1395 | 00**
3	3 cans	Adhesive	8 | 99	26 | 97
4	5 boxes	Tack strips	12 | 79	63 | 95
5	3 gal	Paint per gallon	24 | 98	74 | 94
6	2 gal	Paint per gallon	18 | 99	**37 | 98**
		Subtotal:		
		Tax:		306 | 62
		Shipping:		55 | 00
		TOTAL:		

2. The total direct costs for the Brown house were $119,879.32. Total indirect costs were $4,972.65. The purchase price of the house is $154,000. Calculate the profit.

   $154,000 − $124,852 = $29,148 profit

3. Calculate the percent of profit to the nearest tenth of a percent. Hint: Percent is part of a whole—the whole is the purchase price of the house.

   0.18927 × 100 = 18.927% round to 18.9%

Careers in Construction—Accounting

We talked about the role of a payroll administrator at the end of Chapter 6. A cost administrator is another branch of accounting. Cost administrators track the cost of everything—even the little stuff. That way they can see how much a project really costs. In small companies, a cost administrator may have learned by doing the job, but in big companies, they probably need to have a college degree. They all need to be good at math and very detail-oriented.

Terms

Invoices: Itemized records of material that have been shipped to a customer. An invoice usually states the price of each item in the shipment.

NOTES

Chapter 14
Everyone Has an Angle

Trigonometry is one of the most powerful mathematical tools a student can have. Although the entire study of trigonometry is complex, the most commonly used aspects are fairly simple—chiefly the functions. Help students to apply these functions to any right angle—use the corner of a foundation, the corner of a wall, the peak of a roof, or the corner of a building lot. The point is to get the students to calculate unknown values based on easy-to-obtain known values. Once students start applying these functions, they will see the use in them. Go back to the electrical and plumbing chapters for practical applications in construction.

You can help your students to remember the ratios for sine, cosine, and tangent with the mnemonic "**S**addle **O**ur **H**orses, **C**anter **A**long **H**appily **T**o **O**ther **A**dventures."

Sine = **O**pposite/**H**ypotenuse

Cosine = **A**djacent/**H**ypotenuse

Tangent = **O**pposite/**A**djacent

Note: Trig tables can be found online or accessed using a scientific calculator. They are also provided on the CD supplied with this Instructor's Guide.

NOTES

Chapter 14

Everyone Has an Angle

This chapter reviews trigonometric functions. Trigonometry is the study of triangles, and trigonometric functions are the ratios between the sides of any right triangle. So far you have labeled the sides of a right triangle with the letters a, b, and c. Using these letters, there are six ratios.

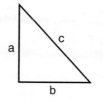

RATIOS	
1. $\frac{a}{b}$	4. $\frac{b}{c}$
2. $\frac{a}{c}$	5. $\frac{c}{a}$
3. $\frac{b}{a}$	6. $\frac{c}{b}$

For trigonometry, you need to learn some new names for the sides. First, in a right triangle there is always a 90-degree angle.

The side opposite the 90-degree angle is called the hypotenuse. The hypotenuse is always the longest side in a right triangle.

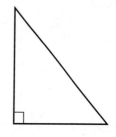

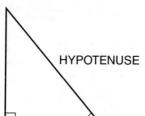

1. HYPOTENUSE IS THE LONGEST SIDE AND IS <u>ALWAYS</u> OPPOSITE THE RIGHT ANGLE.

Because the angle will determine the lengths of the sides, you need to look at the triangle from the point of view of an angle. The side opposite the angle is called the opposite side.

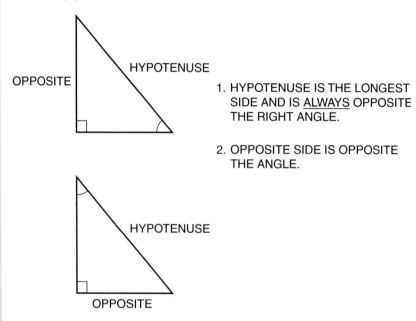

The side next to the angle is called the adjacent side. (The word adjacent means next to or closest.)

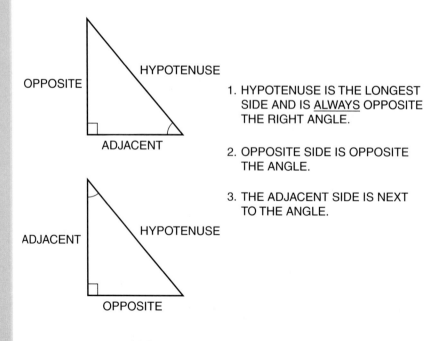

Practice Problems 14-1

It's important to be able to label the sides of the right triangle correctly. Label the sides of the triangles drawn below. Be sure you look at the angle and label the side opposite the angle with the letter O for opposite, and then label the side next to the angle with the letter A for adjacent.

1. A. 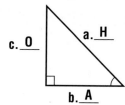 B.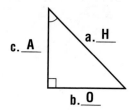

2. A. 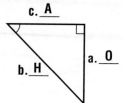 B.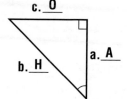

3. A. 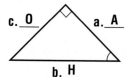 B.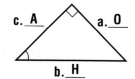

- 14.3 -

4. A.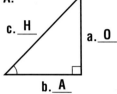

   B.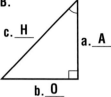

5. A.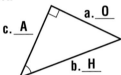

   B.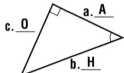

Using the new names for the sides, there are still six ratios.

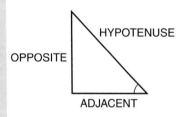

1. $\dfrac{\text{OPPOSITE}}{\text{HYPOTENUSE}}$ 2. $\dfrac{\text{HYPOTENUSE}}{\text{OPPOSITE}}$

3. $\dfrac{\text{ADJACENT}}{\text{HYPOTENUSE}}$ 4. $\dfrac{\text{HYPOTENUSE}}{\text{ADJACENT}}$

5. $\dfrac{\text{OPPOSITE}}{\text{ADJACENT}}$ 6. $\dfrac{\text{ADJACENT}}{\text{OPPOSITE}}$

These ratios have names.

SINE = $\dfrac{\text{OPPOSITE}}{\text{HYPOTENUSE}}$ COSECANT = $\dfrac{\text{HYPOTENUSE}}{\text{OPPOSITE}}$

COSINE = $\dfrac{\text{ADJACENT}}{\text{HYPOTENUSE}}$ SECANT = $\dfrac{\text{HYPOTENUSE}}{\text{ADJACENT}}$

TANGENT = $\dfrac{\text{OPPOSITE}}{\text{ADJACENT}}$ COTANGENT = $\dfrac{\text{ADJACENT}}{\text{OPPOSITE}}$

A trigonometric function table lists the ratios of all angles, three of which are shown in the table below. A trigonometric function table is a handy tool to have because someone else has already figured out the ratios of each angle. These functions are used so often that they are stored in the permanent memory of many calculators.

Deg	Sine	Cosine	Tangent	Cosecant	Secant	Cotangent
30	0.5000	0.8660	0.5774	2.0000	1.1547	1.7321
45	0.7071	0.7071	1.0000	1.4142	1.4142	1.0000
60	0.8660	0.5000	1.7321	1.1547	2.0000	0.5774

Let's apply these functions to a 45-degree right triangle.

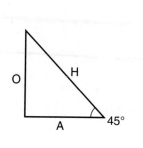

$$\text{SINE } 45° = \frac{O}{H}$$

$$\text{COSINE } 45° = \frac{A}{H}$$

$$\text{TANGENT } 45° = \frac{O}{A}$$

$$\text{COSECANT } 45° = \frac{H}{O}$$

$$\text{SECANT } 45° = \frac{H}{A}$$

$$\text{COTANGENT } 45° = \frac{A}{O}$$

Insert the values from the table.

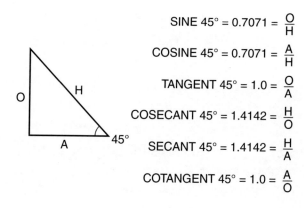

$$\text{SINE } 45° = 0.7071 = \frac{O}{H}$$

$$\text{COSINE } 45° = 0.7071 = \frac{A}{H}$$

$$\text{TANGENT } 45° = 1.0 = \frac{O}{A}$$

$$\text{COSECANT } 45° = 1.4142 = \frac{H}{O}$$

$$\text{SECANT } 45° = 1.4142 = \frac{H}{A}$$

$$\text{COTANGENT } 45° = 1.0 = \frac{A}{O}$$

One more piece of information is needed. You need to know the length of at least one side of the triangle. In this triangle, the adjacent side is equal to 3 inches. Replace any letter A with 3 inches.

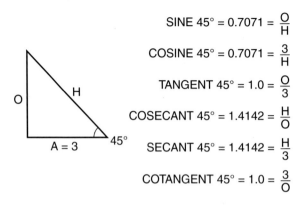

$$\text{SINE } 45° = 0.7071 = \frac{O}{H}$$

$$\text{COSINE } 45° = 0.7071 = \frac{3}{H}$$

$$\text{TANGENT } 45° = 1.0 = \frac{O}{3}$$

$$\text{COSECANT } 45° = 1.4142 = \frac{H}{O}$$

$$\text{SECANT } 45° = 1.4142 = \frac{H}{3}$$

$$\text{COTANGENT } 45° = 1.0 = \frac{3}{O}$$

Finally, solve for the opposite and hypotenuse.

$\text{SINE } 45° = 0.7071 = \frac{O}{H}$ CAN'T SOLVE

$\text{COSINE } 45° = 0.7071 = \frac{3"}{H}$
$= 0.7071H = 3"$ (MULTIPLY BOTH SIDES BY H)
$H = 4.243"$ (DIVIDE BOTH SIDES BY 0.7071)

$\text{TANGENT } 45° = 1.0 = \frac{O}{3"}$
$3" = O$ (MULTIPLY BOTH SIDES BY 3)

$\text{COSECANT } 45° = 1.4142 = \frac{H}{O}$ CAN'T SOLVE

$\text{SECANT } 45° = 1.4142 = \frac{H}{3"}$
$4.243" = H$ (MULTIPLY BOTH SIDES BY 3)

$\text{COTANGENT } 45° = 1.0 = \frac{3"}{O}$
$O = 3"$ (MULTIPLY BOTH SIDES BY O)

Notice that solving the equations for tangent and cotangent confirmed the length of the opposite side, and the equations for cosine and secant did the same for the hypotenuse. At first, you can't solve for the sine or the cosecant, because you don't know the length of the hypotenuse or the opposite side. Once you use one of the other functions to calculate the length of either side, you can go back and solve for the sine and cosecant.

$$\text{SINE } 45° = 0.7071 = \frac{O}{H}$$

$$0.7071 = \frac{3"}{H} \quad \text{(MULTIPLY BOTH SIDES BY H AND DIVIDE BOTH SIDES BY 0.7071)}$$

$$H = 4.243"$$

$$\text{COSECANT } 45° = 1.4142 = \frac{H}{3"}$$

$$4.243" = H \quad \text{(MULTIPLY BOTH SIDES BY 3)}$$

Since this is a 45-degree right triangle, you know that the opposite and adjacent sides must be equal, so you could have used the Pythagorean theorem to solve for the length of the hypotenuse. Whether you use trigonometric functions or the Pythagorean theorem, the length of side c, which is also called the hypotenuse, is the same—4.23 inches.

$$a^2 + b^2 = c^2$$
$$3^2 + 3^2 = c^2$$
$$9 + 9 = c^2$$
$$\sqrt{18} = c^2$$
$$18 = c$$
$$= 4.24$$

Try again, but this time use a 60-degree right triangle with an adjacent side equaling 1 inch.

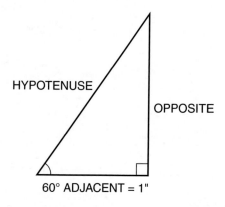

HYPOTENUSE
OPPOSITE
60° ADJACENT = 1"

Start by looking up the ratios for a 60-degree angle in the table shown previously. Then fill in the information you know (the adjacent side of the triangle is 1 inch). Solve for any unknowns. Be sure to go back and solve for the sine and cosecant equation when you have more information.

SINE 60° = 0.8660 = $\frac{O}{H}$ CAN'T SOLVE

$\quad\quad$ 0.8660 = $\frac{O}{2"}$ (MULTIPLY BOTH SIDES BY 2)

$\quad\quad$ 1.732" = O

COSINE 60° = 0.5 = $\frac{1"}{H}$

$\quad\quad$ 0.5H = 1"   (MULTIPLY BOTH SIDES BY H)
$\quad\quad$ H = 2"   (DIVIDE BOTH SIDES BY 0.5)

TANGENT 60° = 1.7321 = $\frac{O}{1"}$

$\quad\quad$ 1.7321" = O   (DIVIDE BOTH SIDES BY 1)

COSECANT 60° = 1.1547 = $\frac{H}{O}$ CAN'T SOLVE

$\quad\quad$ 0.8660 = $\frac{H}{1.732"}$ (MULTIPLY BOTH SIDES BY 1.732)

$\quad\quad$ 2" = H   (ROUNDED UP FROM 1.999404)

SECANT 60° = 2 = $\frac{H}{1}$

$\quad\quad$ 2" = H   (DIVIDE BOTH SIDES BY 1)

COTANGENT 60° = 0.5774 = $\frac{1}{O}$

$\quad\quad$ 0.5774O = 1   (MULTIPLY BOTH SIDES BY O)
$\quad\quad$ O = 1.732"   (DIVIDE BOTH SIDES BY 0.5774)

Chapter 14 Review Problems

1. Find the length of the hypotenuse for the following right triangle.

   Hints:
   First, label the sides opposite, adjacent, and hypotenuse. Second, decide whether to use sine, cosine, tangent, cosecant, secant, or cotangent to find the hypotenuse. Finally, solve for H.

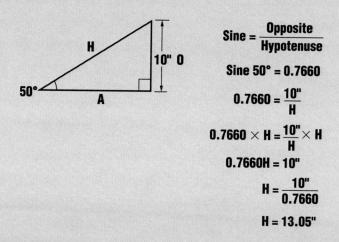

$$\text{Sine} = \frac{\text{Opposite}}{\text{Hypotenuse}}$$

$$\text{Sine } 50° = 0.7660$$

$$0.7660 = \frac{10"}{H}$$

$$0.7660 \times H = \frac{10"}{H} \times H$$

$$0.7660 H = 10"$$

$$H = \frac{10"}{0.7660}$$

$$H = 13.05"$$

2. Find the length of the adjacent side for the following right triangle.

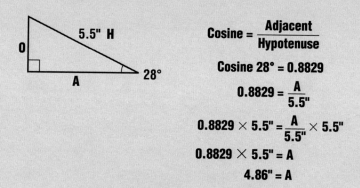

$$\text{Cosine} = \frac{\text{Adjacent}}{\text{Hypotenuse}}$$

$$\text{Cosine } 28° = 0.8829$$

$$0.8829 = \frac{A}{5.5"}$$

$$0.8829 \times 5.5" = \frac{A}{5.5"} \times 5.5"$$

$$0.8829 \times 5.5" = A$$

$$4.86" = A$$

3. Find the length of the opposite side for the following right triangle.

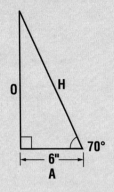

Tangent = $\dfrac{\text{Opposite}}{\text{Adjacent}}$

Tan 70° = 2.7475

$2.7475 = \dfrac{O}{6"}$

$2.7475 \times 6" = \dfrac{O}{6"} \times 6"$

2.7475 × 6" = O

16.485" = O

Epilogue

All of Mr. Whyte's students graduated from high school. But not all of them liked working in construction. Sandy, Phil, and Tom loved it. Mr. Whyte convinced the three of them to try college, so after graduation they each enrolled in the local community college. Sandy goes full time to become a plumber, but Tom and Phil go part time at night. Tom's in the electrician program and Phil's in carpentry. Both of them work for Mr. Conrad at JM Custom Homes during the day. They had to start out on the roofing crew, but they didn't mind. Tom is hoping to become an electrician's helper soon.

Al finally bought Mr. Brown's old red pickup. Mr. Brown let Al work off the cost by putting up a fence to keep the neighbor's cows off the property and doing work around the house for Mrs. Brown. Mr. Brown encouraged Al to apply for scholarships and grants to go to the university. Al is now attending a four-year college to become an architect. It's hard going to school during the day and working as a handyman at a local apartment complex at night, but Al thinks it will be worth it.

Jorge decided he liked accounting. He's currently attending the community college and hopes to be accepted into the university next year.

Travis decided he wasn't too crazy about construction, but he was captivated by surveying. He wasn't interested in going to college, but he worked for Mr. Burke during the summer who convinced Travis to attend an NCCER Site Layout course in Georgia.

Olivia decided she wasn't cut out for construction, but she found out that she had a lot of patience in dealing with the public. She is now working as a sales representative for JM Custom Homes. She explains the construction process to the homeowner and helps to guide them through all phases of construction. She often visits building sites with the homeowner to answer questions about carpeting, paint, and appliances. The construction workers really like having Olivia in that role, because it saves them from having to answer a lot of questions.

NOTES